AF683104

# SOLUTIONS MANUAL TO ACCOMPANY

# QUANTUM CHEMISTRY

Donald A. McQuarrie
Department of Chemistry
University of California, Davis

12 13 23 38

UNIVERSITY SCIENCE BOOKS
Mill Valley, California

University Science Books
20 Edgehill Road, Mill Valley, CA 94941

Copyright © 1984 by University Science Books.
Reproduction or translation of any part of this work beyond that permitted by Sections 107 or 108 of the 1976 United States Copyright Act without permission of the copyright owner is unlawful. Requests for permission or further information should be addressed to the Permissions Department, University Science Books.

Library of Congress Catalog Card Number 84-051961
ISBN 0-935702-16-4

PRINTED IN THE UNITED STATES OF AMERICA
5 4 3 2 1

# PREFACE

This Solutions Manual is meant to accompany my text "Quantum Chemistry" and contains detailed solutions to each of the 360 problems in the text. The only way to learn a subject like quantum chemistry is to do many problems, and the detailed solutions presented here can serve as a valuable aid for that purpose, but only if it is used correctly. This manual is no substitute for working the problems, and should not be consulted before a sincere effort has been made to do any problem.

Better stated,

"But be ye doers of the word, and not hearers only, deceiving your own selves."

James i,22

I would like to thank Louise Hope for typing the manuscript and preparing it for copy.

## CONTENTS

# CHAPTER ONE

# HISTORICAL BACKGROUND

**1-1.** Use the equation $\nu\lambda = c$.

$$\nu = \frac{c}{\lambda} = \frac{2.998 \times 10^{8}\ \text{m}\cdot\text{s}^{-1}}{2000 \times 10^{-10}\ \text{m}} = 1.50 \times 10^{15}\ \text{s}^{-1}$$

Use the equation $\bar{\nu} = 1/\lambda$.

$$\bar{\nu} = \frac{1}{\lambda} = \frac{1}{2000 \times 10^{-10}\ \text{m}} = 5.00 \times 10^{6}\ \text{m}^{-1}$$

$$= 5.00 \times 10^{4}\ \text{cm}^{-1}$$

$$E = h\nu = (6.626 \times 10^{-34}\ \text{J}\cdot\text{s})(1.50 \times 10^{15}\ \text{s}^{-1})$$

$$= 9.94 \times 10^{-19}\ \text{J}$$

**1-2.** We are given that $\bar{\nu} = 10^{3}\ \text{cm}^{-1} = 1/\lambda$ and so $\lambda = 10^{-3}\ \text{cm} = 10^{-5}\ \text{m}$. Therefore

$$\nu = \frac{c}{\lambda} = \frac{3.00 \times 10^{8}\ \text{m}\cdot\text{s}^{-1}}{10^{-5}\ \text{m}} = 3.0 \times 10^{13}\ \text{s}^{-1}$$

and

$$E = h\nu = (6.626 \times 10^{-34}\ \text{J}\cdot\text{s})(3.0 \times 10^{13}\ \text{s}^{-1})$$

$$= 2.0 \times 10^{-20}\ \text{J}$$

**1-3.**

$$\bar{\nu} = \frac{\nu}{c} = \frac{2.0 \times 10^{10}\ \text{s}^{-1}}{3.0 \times 10^{8}\ \text{m}\cdot\text{s}^{-1}} = 67\ \text{m}^{-1} = 0.67\ \text{cm}^{-1}$$

$$\lambda = \frac{c}{\nu} = \frac{3.0 \times 10^{8}\ \text{m}\cdot\text{s}^{-1}}{2.0 \times 10^{10}\ \text{s}^{-1}} = 1.5 \times 10^{-2}\ \text{m}$$

$$E = h\nu = (6.626 \times 10^{-34}\ \text{J}\cdot\text{s})(2.0 \times 10^{10}\ \text{s}^{-1})$$

$$= 1.32 \times 10^{-23}\ \text{J}$$

**1-4.** Equation 1-2 is

$$\rho(\nu,T)d\nu = \frac{8\pi h\nu^{3}}{c^{3}}\,\frac{d\nu}{e^{h\nu/kT} - 1}$$

We now use the expansion $e^{x} = 1 + x + x^{2}/2 + \cdots$

As $\nu \to 0$,

$$e^{h\nu/kT} = 1 + \frac{h\nu}{kT} + \cdots$$

and so

$$\rho(\nu,T)\,d\nu = \frac{8\pi h\nu^3}{c^3}\frac{d\nu}{h\nu/kT} = \frac{8\pi kT}{c^3}\nu^2 d\nu$$

**1-5.** We start with Eq. 1-3

$$\rho'(\lambda,T) = \frac{8\pi hc}{\lambda^5}\frac{1}{e^{hc/\lambda kT} - 1}$$

We differentiate with respect to $\lambda$.

$$\frac{d\rho'}{d\lambda} = \frac{8\pi hc}{(e^{hc/\lambda kT} - 1)}\left[-\frac{5}{\lambda^6} + \frac{\frac{hc}{\lambda^2 kT}e^{hc/\lambda kT}}{\lambda^5(e^{hc/\lambda kT} - 1)}\right] = 0$$

or

$$5 = \frac{hc}{\lambda kT}\frac{e^{hc/\lambda kT}}{(e^{hc/\lambda kT} - 1)}$$

Let $x = hc/\lambda kT$ and rearrange slightly to get

$$e^{-x} + \frac{x}{5} = 1$$

By trial and error, we find that $x = 4.965$

**1-6.** We use Eq. 1-4

$$\lambda_{max}T = 2.90 \times 10^{-3}\ \text{m}\cdot\text{K}$$

Then

a) $\lambda_{max} = 9.67 \times 10^{-6}$ m at 300 K

b) $\lambda_{max} = 9.67 \times 10^{-7}$ m at 3000 K

c) $\lambda_{max} = 2.90 \times 10^{-7}$ m at 10,000 K

**1-7.** We use Eq. 1-4

$$\lambda_{max}T = 2.90 \times 10^{-3}\ \text{m}\cdot\text{K}$$

Then

$$T = \frac{2.90 \times 10^{-3}\ \text{m}\cdot\text{K}}{2600 \times 10^{-10}\ \text{m}} = 1.1 \times 10^4\ \text{K}$$

**1-8.** We use Eq. 1-4 to get

$$\lambda_{max} = \frac{2.90 \times 10^{-3}\ \text{m}\cdot\text{K}}{10^{7}\ \text{K}} = 2.90 \times 10^{-10}\ \text{m}$$

$$= 2.90\ \text{Å (X-rays)}$$

**1-9.** We start with Eq. 1-9.

$$\sigma = \frac{2\pi^5 k^4}{15h^3c^2}$$

$$= \frac{2\pi^5(1.381 \times 10^{-23}\ \text{J}\cdot\text{K}^{-1})^4}{15(6.626 \times 10^{-34}\ \text{J}\cdot\text{s})^3(2.998 \times 10^{8}\ \text{m}\cdot\text{s}^{-1})^2}$$

$$= 5.670 \times 10^{-8}\ \text{J}\cdot\text{m}^{-2}\cdot\text{K}^{-4}\cdot\text{s}^{-1}$$

**1-10.** Start with

$$\lambda_{max}T = \frac{hc}{4.965\ k} = 2.90 \times 10^{-3}\ \text{m}\cdot\text{K}$$

and

$$\sigma = \frac{2\pi^5 k^4}{15h^3c^2} = 5.6697 \times 10^{-8}\ \text{J}\cdot\text{m}^{-2}\cdot\text{K}^{-4}\cdot\text{s}^{-1}$$

Solve for h and k to obtain

$$h = 6.63 \times 10^{-34}\ \text{J}\cdot\text{s}$$

$$k = 1.38 \times 10^{-23}\ \text{J}\cdot\text{K}^{-1}\cdot\text{molecule}^{-1}$$

Now use the relation

$$N_0 = \frac{R}{k} = \frac{8.314\ \text{J}\cdot\text{K}^{-1}\cdot\text{mol}^{-1}}{1.38 \times 10^{-23}\ \text{J}\cdot\text{K}^{-1}\cdot\text{molecule}^{-1}}$$

$$= 6.02 \times 10^{23}\ \text{molecule}\cdot\text{mol}^{-1}$$

to determine Avogadro's number.

**1-11.** We start with Eq. 1-11.

$$\frac{1}{2}mv^2 = h\nu - \phi$$

$$= (6.626 \times 10^{-34}\ \text{J}\cdot\text{s})\left[\frac{2.998 \times 10^{8}\ \text{m}\cdot\text{s}^{-1}}{2000 \times 10^{-10}\ \text{m}}\right]$$

$$- (4.40\ \text{eV})(1.602 \times 10^{-19}\ \text{J}\cdot\text{eV}^{-1})$$

$$= 2.88 \times 10^{-19}\ \text{J}$$

$$= 1.80\ \text{eV}$$

The stopping potential is 1.80 V.

**1-12.** We use Eq. 1-13.

$$-eV_s = h\nu - \phi = \frac{hc}{\lambda} - \phi$$

Solving for $\phi$, we have

$$\phi = \frac{hc}{\lambda} + eV_s$$

$$= \frac{(6.626\times10^{-34}\ \text{J}\cdot\text{s})(3.00\times10^{8}\ \text{m}\cdot\text{s}^{-1})}{230\times10^{-9}\ \text{m}} - 0.80\text{eV}$$

$$= (8.643\times10^{-19}\ \text{J})\left[\frac{\text{eV}}{1.602\times10^{-19}\ \text{J}}\right] - 0.80\ \text{eV}$$

$$= 4.59\ \text{eV}$$

The threshold frequency is given by

$$h\nu_0 = \phi$$

or

$$\nu_0 = \frac{\phi}{h} = \frac{(4.59\ \text{eV})(1.602\times10^{-19}\ \text{J}\cdot\text{eV}^{-1})}{(6.626\times10^{-34}\ \text{J}\cdot\text{s})}$$

$$= 1.11\times10^{15}\ \text{s}^{-1}$$

**1-13.** The frequencies of the lines in the Lyman series are given by

$$\bar{\nu} = (109{,}680\ \text{cm}^{-1})\left[\frac{1}{1} - \frac{1}{n^2}\right] \qquad n = 2,3,\ldots$$

When n = 2,

$$\bar{\nu} = 82{,}260\ \text{cm}^{-1} \qquad \text{or} \qquad \lambda = \frac{1}{\bar{\nu}} = 1216\ \text{Å}$$

When n is very large,

$$\bar{\nu} = 109{,}680\ \text{cm}^{-1} \qquad \text{or} \qquad \lambda = \frac{1}{\bar{\nu}} = 912\ \text{Å}$$

For the Balmer series we use the equation

$$\bar{\nu} = (109{,}680\ \text{cm}^{-1})\left[\frac{1}{4} - \frac{1}{n^2}\right] \qquad n = 3,4,\ldots$$

When n=3,

$$\bar{\nu} = (109{,}680\ \text{cm}^{-1})\left[\frac{1}{4} - \frac{1}{9}\right] = 15{,}233\ \text{cm}^{-1}$$

$$\text{or } \lambda = \frac{1}{\bar{\nu}} = 6563\ \text{Å}$$

and when n is very large

$$\bar{\nu} = (109{,}680\ \text{cm}^{-1})\left[\frac{1}{4}\right] = 27{,}420\ \text{cm}^{-1}$$

$$\text{or } \lambda = \frac{1}{\bar{\nu}} = 3647\ \text{Å}$$

For the Paschen series we use

$$\bar{\nu} = (109{,}680\ \text{cm}^{-1})\left[\frac{1}{9} - \frac{1}{n^2}\right] \qquad n = 4,5,\ldots$$

When n=4,

$$\bar{\nu} = (109{,}680\ \text{cm}^{-1})\left[\frac{1}{9} - \frac{1}{16}\right] = 5331\ \text{cm}^{-1}$$

$$\text{or } \lambda = \frac{1}{\bar{\nu}} = 18{,}760\ \text{Å}$$

and when n is very large

$$\bar{\nu} = (109{,}680\ \text{cm}^{-1})\left[\frac{1}{9}\right] = 12{,}187\ \text{cm}^{-1}$$

$$\text{or } \lambda = \frac{1}{\bar{\nu}} = 8206\ \text{Å}$$

**1-14.** The units of Planck's constant are J·s, and so

$$\text{J·s} = (\text{kg·m}^2\text{·s}^{-2})(\text{s}) = \text{kg·m}^2\text{·s}^{-1}$$

**1-15.** $R = 109{,}680\ \text{cm}^{-1}$

Therefore

$$\lambda = \frac{1}{109{,}680\ \text{cm}^{-1}} = 9.117 \times 10^{-8}\ \text{m}$$

and

$$\nu = \frac{c}{\lambda} = \frac{2.998 \times 10^{8}\ \text{m·s}^{-1}}{9.117 \times 10^{-8}\ \text{m}} = 3.288 \times 10^{15}\ \text{s}^{-1}$$

and

$$E = h\nu = (6.626 \times 10^{-34}\ \text{J·s})(3.288 \times 10^{15}\ \text{s}^{-1})$$
$$= 2.179 \times 10^{-18}\ \text{J}$$

This is the energy corresponding to one photon. Per mole we have

$$E = (2.179 \times 10^{-18}\ \text{J})(6.022 \times 10^{23}\ \text{mol}^{-1})$$

$$= 1.312 \times 10^{6}\ \text{J}\cdot\text{mol}^{-1}$$

$$= 1.312 \times 10^{3}\ \text{kJ}\cdot\text{mol}^{-1}$$

Lastly

$$E = (2.179 \times 10^{-18}\ \text{J})\left[\frac{1\text{eV}}{1.602 \times 10^{-19}\ \text{J}}\right]$$

$$= 13.60\ \text{eV}$$

**1-16.** The reduced mass is given by Eq. 1-40

a) $$\mu = \frac{m_1 m_2}{m_1 + m_2}$$

$$= \left[\frac{(14.0)(14.0)}{28.0}\text{amu}\right](1.661 \times 10^{-27}\ \text{kg}\cdot\text{amu}^{-1})$$

$$= 1.162 \times 10^{-26}\ \text{kg}$$

b) $$\mu = \left[\frac{(1.008)(34.97)}{35.98}\text{amu}\right](1.661 \times 10^{-27}\ \text{kg}\cdot\text{amu}^{-1})$$

$$= 1.627 \times 10^{-27}\ \text{kg}$$

**1-17.** Solve the expression given for I and find that

$$I = \frac{h}{4\pi^2 B} = \frac{6.626 \times 10^{-34}\ \text{J}\cdot\text{s}}{4\pi^2(3.13 \times 10^{11}\ \text{s}^{-1})}$$

$$= 5.362 \times 10^{-47}\ \text{kg}\cdot\text{m}^2$$

The reduced mass is given by (see problem 1-16b)

$$\mu = \frac{m_1 m_2}{m_1+m_2} = 1.627 \times 10^{-27}\ \text{kg}$$

and so

$$r = \left[\frac{I}{\mu}\right]^{1/2} = \left[\frac{5.362 \times 10^{-47}\ \text{kg}\cdot\text{m}^2}{1.627 \times 10^{-27}\ \text{kg}}\right]^{1/2}$$

$$= 1.815 \times 10^{-10}\ \text{m}$$

**1-18.**

$$R_H = \frac{\mu e^4}{8\varepsilon_0 c h^3}$$

$$= \frac{(9.1046 \times 10^{-31}\ \text{kg})(1.60219 \times 10^{-19}\ \text{C})^4}{(8)(8.854188 \times 10^{-12}\ \text{C}^2\cdot\text{s}^2\cdot\text{kg}^{-1}\cdot\text{m}^{-3})^2 \times (2.997925 \times 10^{8}\ \text{m}\cdot\text{s}^{-1}) \times (6.62618 \times 10^{-34}\ \text{J}\cdot\text{s})^3}$$

$$= 1.0968 \times 10^7 \text{ m}^{-1}$$

**1-19.** The two key equations in the Bohr theory are

$$mvr = n\hbar \qquad (1\text{-}23)$$

and

$$\frac{e^2}{4\pi\varepsilon_o r^2} = \frac{mv^2}{r} \qquad (1\text{-}22)$$

For a nucleus of charge Ze, Eq. 1-22 becomes

$$\frac{Ze^2}{4\pi\varepsilon_o r^2} = \frac{mv^2}{r}$$

Eliminate v between this equation and Eq. 1-23 to obtain

$$r = \frac{4\pi\varepsilon_o\hbar^2 n^2}{mZe^2}$$

The total energy of the electron is (compare to Eq. 1-27)

$$E = \frac{1}{2}mv^2 - \frac{Ze^2}{4\pi\varepsilon_o r} = \frac{1}{2}\left[\frac{Ze^2}{4\pi\varepsilon_o r}\right] - \frac{Ze^2}{4\pi\varepsilon_o r} = -\frac{Ze^2}{8\pi\varepsilon_o r}$$

If we substitute the above expression for r into this, then we get

$$E = -\frac{Ze^2}{8\pi\varepsilon_o}\,\frac{mZe^2}{4\pi\varepsilon_o\hbar^2 n^2} = -\frac{mZ^2e^4}{8\varepsilon_o^2 h^2 n^2}$$

The corresponding Rydberg formula is

$$\Delta E = \frac{me^4Z^2}{8\varepsilon_o^2h^2}\left[\frac{1}{n_1^2} - \frac{1}{n_2^2}\right]$$

Dividing by hc to convert $\Delta E$ to $\bar{\nu}$ we obtain

$$\bar{\nu} = \frac{me^4Z^2}{8\varepsilon_o^2h^3c}\left[\frac{1}{n_1^2} - \frac{1}{n_2^2}\right]$$

$$= R_H Z^2\left[\frac{1}{n_1^2} - \frac{1}{n_2^2}\right]$$

**1-20.** We use Eq. 1-47, which is derived in the previous problem. We set $n_1 = 4$ and $n_2 = 5, 6, \ldots$

and write

$$\bar{\nu} = \frac{Z^2e^4\mu}{8\varepsilon_0{}^2ch^3}\left[\frac{1}{16} - \frac{1}{n^2}\right] \qquad n = 5,6,\ldots$$

where Z = 2 and

$$\mu = \frac{(9.1096 \times 10^{-31})(6.644 \times 10^{-27})}{9.1096 \times 10^{-31} + 6.644 \times 10^{-27}} \text{ kg}$$

$$= 9.1084 \times 10^{-31} \text{ kg}$$

Consequently the lines in the Pickering series are given by

$$\bar{\nu} = (4)(1.0973 \times 10^7 \text{ m}^{-1})\left[\frac{1}{16} - \frac{1}{n^2}\right]$$

$$= 4.389 \times 10^7\left[\frac{1}{16} - \frac{1}{n^2}\right]\text{m}^{-1}$$

When $n = 5$, $\bar{\nu} = 9.875 \times 10^5 \text{ m}^{-1}$, which is in the infrared region.

**1-21.** We use Eq. 1-47. To find the ionization energy, we set $n_1 = 1$ and ignore the $1/n_2{}^2$ term and write

$$\text{IE} = Z^2(13.60 \text{ eV}) = 4(13.60 \text{ eV}) = 54.4 \text{ eV}$$

We have used the fact that one Rydberg is equal to 13.60 eV (Problem 1-15).

**1-22.** The reduced mass of this system is given by

$$\mu = \frac{(207m_e)(m_p)}{207m_e + m_p} = 1.695 \times 10^{-28} \text{ kg}$$

where $m_e$ is the mass of an electron and $m_p$ is the mass of a proton.

The radius of the first Bohr orbit is given by

$$r = \frac{4\pi\varepsilon_0\hbar^2}{\mu e^2} = 0.284 \text{ pm} \qquad (n = 1)$$

and the allowed energies are given by

$$E_n = -\frac{\mu e^4}{8\varepsilon_0{}^2h^2n^2} = -4.06 \times 10^{-16} \text{ J}$$

The energy associated with the n = 1 to n = 2 transition is

$$\Delta E = (4.06 \times 10^{-16}\ \text{J})\left[\frac{1}{1} - \frac{1}{4}\right] = 3.04 \times 10^{-16}\ \text{J}$$

and the frequency is

$$\nu = \frac{\Delta E}{h} = 4.60 \times 10^{17}\ \text{s}^{-1}$$

**1-23.** The deBroglie wave length of a moving particle is given by (Eq. 1-49)

$$\lambda = \frac{h}{p} = \frac{h}{mv} = \frac{h}{(2mE)^{1/2}}$$

(a) For an electron with a kinetic energy of 100 eV,

$$\lambda = \frac{6.626 \times 10^{-34}\ \text{J}\cdot\text{s}}{[2(9.110 \times 10^{-31}\ \text{kg})(100\ \text{eV}) \times (1.602 \times 10^{-19}\ \text{J}\cdot\text{eV}^{-1})]^{1/2}}$$

$$= 1.23 \times 10^{-10}\ \text{m}$$

(b) For a proton with a kinetic energy of $10^5$ eV,

$$\lambda = \frac{6.626 \times 10^{-34}\ \text{J}\cdot\text{s}}{[2(1.673 \times 10^{-27}\ \text{kg})(10^5\ \text{eV})(1.602 \times 10^{-19}\ \text{J}\cdot\text{eV}^{-1})]^{1/2}}$$

$$= 9.05 \times 10^{-14}\ \text{m}$$

(c) For an electron in the first Bohr orbit of a hydrogen atom

$$\lambda = \frac{h}{mv} = \frac{6.626 \times 10^{-34}\ \text{J}\cdot\text{s}}{(9.110 \times 10^{-31}\ \text{kg})(2.19 \times 10^6\ \text{m}\cdot\text{s}^{-1})}$$

$$= 3.32 \times 10^{-10}\ \text{m}$$

We have used the result of Example 1-14, where it is shown that the speed of an electron in the first Bohr orbit of a hydrogen atom is $2.19 \times 10^6\ \text{m}\cdot\text{s}^{-1}$

**1-24.** This problem is similar to Problem 1-23.

(a) $E = 100\ \text{eV} = 1.602 \times 10^{-17}\ \text{J}$

$$\lambda = \frac{h}{(2mE)^{1/2}}$$

$$= \frac{6.626 \times 10^{-34}\ \text{J·s}}{[(2)(9.110 \times 10^{-31}\ \text{kg})(1.602 \times 10^{-17}\ \text{J})]^{1/2}}$$

$$= 1.23 \times 10^{-10}\ \text{m}$$

(b) We are given a value of $\lambda$ and we must determine E.

$$\lambda = 2.00\ \text{Å} = 2.00 \times 10^{-10}\ \text{m}$$

$$E = \frac{p^2}{2m} = \frac{(h/\lambda)^2}{2m}$$

$$= \left[\frac{6.626 \times 10^{-34}\ \text{J·s}}{2.00 \times 10^{-10}\ \text{m}}\right]^2 \frac{1}{(2)(9.110\times10^{-31}\ \text{kg})}$$

$$= 6.02 \times 10^{-18}\ \text{J}$$

**1-25.** We are given $\lambda$ and wish to find E in electron volts.

$$E = \frac{p^2}{2m} = \frac{h^2}{2m\lambda^2} = \frac{(6.626 \times 10^{-34}\ \text{J·s})^2}{2(1.673 \times 10^{-27}\ \text{kg})(10^{-10}\ \text{m})^2}$$

$$= (1.31 \times 10^{-20}\ \text{J})\left[\frac{1\ \text{eV}}{1.602 \times 10^{-19}\ \text{J}}\right]$$

$$= 8.19 \times 10^{-2}\ \text{eV}$$

Thus the potential through which the proton must fall is 81.9 mV

**1-26.** This problem is similar to Problems 1-23 and 1-24.

$$E = (2\ e)(4.0\ \text{V}) = 8.0\ \text{eV} = 1.28 \times 10^{-18}\ \text{J}$$

$$\lambda = \frac{h}{(2mE)^{1/2}}$$

$$= \frac{6.626 \times 10^{-34}\ \text{J·s}}{[(2)(4)(1.66 \times 10^{-27}\ \text{kg})(1.28 \times 10^{-18}\ \text{J})]^{1/2}}$$

$$= 5.06 \times 10^{-12}\ \text{m}$$

**1-27.** We substitute Eq. 1-24 into 1-23 and obtain

$$v = \frac{e^2}{4\pi\varepsilon_0\hbar n} = \frac{e^2}{2\varepsilon_0 hn}$$

The speed for the first few Bohr orbits is given by

$$v = \frac{(1.602 \times 10^{-19}\ \text{C})^2}{2(8.854 \times 10^{-12}\ \text{C}^2\cdot\text{s}^2\cdot\text{kg}^{-1}\cdot\text{m}^{-3}) \times (6.626 \times 10^{-34}\ \text{J}\cdot\text{s})n}$$

$$= \frac{2.19 \times 10^6\ \text{m}\cdot\text{s}^{-1}}{n} \qquad n = 1,2,\ldots$$

**1-28.**

$$K = \int_{v=0}^{v=v} v\, d\left[\frac{m_0 v}{\sqrt{1 - (v^2/c^2)}}\right]$$

Factor out $m_0$ and integrate by parts to obtain

$$K = m_0 \left|\frac{v^2}{\sqrt{1 - (v^2/c^2)}}\right|_{v=0}^{v=v} - m_0 \int_0^v \frac{v'dv'}{\sqrt{1 - (v'^2/c^2)}}$$

$$= \frac{m_0 v^2}{\sqrt{1 - (v^2/c^2)}} - m_0 c \left| - \sqrt{c^2 - v^2}\right|_0^v$$

$$= m_0 \left[\frac{v^2}{\sqrt{1 - (v^2/c^2)}} + c\sqrt{c^2 - v^2} - c^2\right]$$

$$= m_0 \left[\frac{c^2}{\sqrt{1 - (v^2/c^2)}} - c^2\right]$$

$$= m_0 c^2 \left[\frac{1}{\sqrt{1 - (v^2/c^2)}} - 1\right]$$

As $v/c \rightarrow 0$,

$$\frac{1}{\sqrt{1 - (v^2/c^2)}} = \left[1 - \frac{v^2}{c^2}\right]^{-1/2}$$

$$= 1 + \frac{1}{2}\frac{v^2}{c^2} + \frac{3}{8}\left[\frac{v^2}{c^2}\right]^2 + \cdots$$

and so

$$K = m_0 c^2 \left[\frac{1}{2}\frac{v^2}{c^2} + \cdots\right] = \frac{1}{2} m_0 v^2 + \cdots$$

By combining Eqs. (1-53) and (1-58)

$$K = (m - m_0)c^2$$

or

$$K = E - m_0c^2$$

From Eq. (1-58),

$$(K + m_0c^2)^2 = \frac{m_0^2c^4}{1 - (v^2/c^2)}$$

From Eq. (1-54)

$$\frac{v^2}{c^2} = \frac{p^2}{m_0^2c^2 + p^2}$$

and so

$$(K + m_0c^2)^2 = c^2p^2 + m_0^2c^4$$

Using Eq. (1-60) for E, we have

$$E^2 = c^2p^2 + m_0^2c^4$$

or

$$E = [(pc)^2 + (m_0c^2)^2]^{1/2}$$

**1-29.** We draw a figure that is analogous to Figure 1-15, but with $\mathbf{r}_1$ replaced by $\mathbf{v}_1$ and $\mathbf{r}_2$ replaced by $\mathbf{v}_2$. By analogy with Figure 1-15, we write

$$|\Delta v| \approx \Delta s = v\Delta\theta$$

Therefore

$$a = \lim_{\Delta t \to 0} \frac{\Delta v}{\Delta t} = v \lim_{\Delta t \to 0} \frac{\Delta\theta}{\Delta t} = v\omega$$

# CHAPTER TWO

# THE WAVE EQUATION

**2-1.**

a)

$$\frac{d^2y}{dx^2} - 4\frac{dy}{dx} + 3y = 0$$

Substitute $y(x) = e^{\alpha x}$ to obtain

$$\alpha^2 - 4\alpha + 3 = 0$$

or

$$(\alpha - 3)(\alpha - 1) = 0$$

The two roots are $\alpha = 1$ and $\alpha = 3$ and the general solution is

$$y(x) = c_1 e^{3x} + c_2 e^{x}$$

b)

$$\frac{d^2y}{dx^2} + 6\frac{dy}{dx} = 0$$

Thus we find

$$\alpha^2 + 6\alpha = 0$$

or that $\alpha = 0$ and $\alpha = -6$. The general solution is

$$y(x) = c_1 + c_2 e^{-6x}$$

c)

$$\frac{dy}{dx} + 3y = 0$$

$$\alpha + 3 = 0$$

or

$$y(x) = c_1 e^{-3x}$$

d)

$$\frac{d^2y}{dx^2} + 2\frac{dy}{dx} - y = 0$$

$$\alpha^2 + 2\alpha - 1 = 0$$

$$\alpha = -1 \pm \frac{1}{2}(4 + 4)^{1/2} = -1 \pm \sqrt{2}$$

The general solution is

$$y(x) = c_1 \exp(-x + \sqrt{2}x) + c_2 \exp(-x - \sqrt{2}x)$$

$$= e^{-x}(c_1 e^{\sqrt{2}x} + c_2 e^{-\sqrt{2}x})$$

e)

$$\frac{d^2y}{dx^2} - 3\frac{dy}{dx} + 2y = 0$$

$$\alpha^2 - 3\alpha + 2 = 0$$

$$(\alpha - 2)(\alpha - 1) = 0$$

$$y(x) = c_1 e^{2x} + c_1 e^{x}$$

**2-2.** This problem is similar to Problem 2-1, except that specific conditions are imposed on the solutions.

a)

$$\frac{d^2y}{dx^2} - 4y = 0$$

$$\alpha^2 - 4 = 0$$

$$\alpha = \pm 2$$

The general solution is

$$y(x) = c_1 e^{2x} + c_2 e^{-2x}$$

The two conditions on $y(x)$ give

$$y(0) = 2 = c_1 + c_2$$

$$y'(0) = 4 = 2c_1 - 2c_2$$

or $c_1 = 2$ and $c_2 = 0$. The solution is

$$y(x) = 2e^{2x}$$

b)

$$\frac{d^2y}{dx^2} - 5\frac{dy}{dx} + 6y = 0$$

$$\alpha^2 - 5\alpha + 6 = 0$$

$$\alpha = 2, 3$$

The general solution is

$$y(x) = c_1 e^{2x} + c_2 e^{3x}$$

The two conditions give

$$y(0) = -1 = c_1 + c_2$$

$$y'(0) = 0 = 2c_1 + 3c_2$$

or $c_1 = -3$ and $c_2 = 2$. The solution is

$$y(x) = 2e^{3x} - 3e^{2x}$$

c)

$$\frac{dy}{dx} - 2y = 0$$

$$\alpha - 2 = 0$$

The general solution is

$$y(x) = ce^{2x}$$

Because $y(0) = 2$, we must have $c = 2$.

**2-3.** We shall prove that $\cos\omega t$ oscillates with a frequency $\nu = \omega/2\pi$.

Replace t by t plus an integral multiple of the period of vibration $\tau = 1/\nu = 2\pi/\omega$. Then

$$\cos[\omega(t + 2\pi n/\omega)] = \cos(\omega t + 2\pi n)$$
$$= \cos\omega t \cos 2\pi n - \sin\omega t \sin 2\pi n$$
$$= \cos\omega t$$

Thus $\cos\omega t$ repeats for integral multiples of $\tau = 2\pi/\omega$, and so the frequency is $1/\tau = \omega/2\pi$. The same argument applies to $\sin\omega t$, and hence to $A\cos\omega t + B\sin\omega t$.

**2-4.**

a)

$$\frac{d^2x}{dt^2} + \omega^2 x = 0$$

From Example 2-4, the general solution to this equation is

$$x(t) = A\cos\omega t + B\sin\omega t$$

The two conditions give

$$x(0) = 0 = A$$

$$x'(0) = v_0 = B\omega$$

or $A = 0$ and $B = v_0/\omega$.

b) The general solution is

$$x(t) = A\cos\omega t + B\sin\omega t$$

The two conditions give

$$x(0) = A$$

$$x'(0) = v_0 = B\omega$$

and so

$$x(t) = A\cos\omega t + \frac{v_0}{\omega}\sin\omega t$$

**2-5.** Let's start with

$$x(t) = A\sin(\omega t + \phi) = A\cos\phi \sin\omega t + A\sin\phi \cos\omega t$$

Comparing this with

$$x(t) = c_1\cos\omega t + c_2\sin\omega t$$

we see that

$$c_1 = A\sin\phi \qquad c_2 = A\cos\phi$$

Similarly,

$$x(t) = B\cos(\omega t + \Psi) = B\cos\Psi \cos\omega t - B\sin\Psi \sin\omega t$$

Comparing this result with $x(t) = c_1\cos\omega t + c_2\sin\omega t$ shows that

$$c_1 = B\cos\Psi \qquad c_2 = -B\sin\Psi$$

To show that $A\sin(\omega t + \phi)$ oscillates with frequency $\omega/2\pi$, replace t with $t + n\tau$, where $\tau = 2\pi/\omega$ is the period of oscillation. This substitution gives

$$\begin{aligned} A\sin(\omega t + \phi) &= A\sin[\omega(t + 2\pi n/\omega) + \phi] \\ &= A\sin[(\omega t + \phi) + 2\pi n] \\ &= A\sin(\omega t + \phi)\cos 2\pi n \\ &\qquad + A\cos(\omega t + \phi)\sin 2\pi n \\ &= A\sin(\omega t + \phi) \end{aligned}$$

Thus, $\tau = 2\pi/\omega$ is indeed the period of oscillation and $\omega/2\pi$ is the frequency.

**2-6.** a) We shall write out the steps in outline.

$$\alpha^2 + 2\alpha + 2 = 0$$

$$\alpha = -1 \pm i$$

$$y(x) = e^{-x}(c_1\cos x + c_2\sin x)$$

b) $\alpha^2 - 6\alpha + 25 = 0$

$\alpha = 3 \pm 4i$

$y(x) = e^{3x}(c_1\cos 4x + c_2\sin 4x)$

c) $\alpha^2 + 2\beta\alpha + (\beta^2 + \omega^2) = 0$

$\alpha = -\beta \pm i\omega$

$y(x) = e^{-\beta x}(c_1\cos\omega x + c_2\sin\omega x)$

d) $\alpha^2 + 4\alpha + 5 = 0$

$\alpha = -2 \pm i$

$y(x) = e^{-2x}(c_1\cos x + c_2\sin x)$

The two conditions give

$y(0) = 1 = c_1$

$y'(0) = -3 = -2c_1 + c_2$

or $c_1 = 1$ and $c_2 = -1$.

Thus $y(x) = e^{-2x}(\cos x - \sin x)$

**2-7.** Start with

$$m\frac{d^2\xi}{dt^2} + k\xi = 0$$

whose solution is

$\xi(t) = A\sin\omega t + B\cos\omega t$

where $\omega = (k/m)^{1/2}$.

The initial condition $x = x_0$ at $t = 0$ corresponds to $\xi(0) = 0$.

Thus we have

$\xi(0) = 0 = B$

The other initial condition is $\dot{\xi}(0) = v_0$, or

$$\frac{d\xi}{dt} = v_0 = A\omega\cos\omega t \quad \text{at } t = 0$$

or

$A\omega = v_0$

The complete solution is

$$\xi(t) = \frac{v_0}{\omega}\sin\omega t = v_0\left[\frac{m}{k}\right]^{1/2}\sin\left[\left[\frac{k}{m}\right]^{1/2}t\right]$$

This solution represents oscillatory motion of frequency $(k/m)^{1/2}/2\pi$ (see Problem 2-5) and amplitude $v_0(m/k)^{1/2}$.

**2-8.** Newton's law states that

$$m\frac{d^2\xi}{dt^2} = \text{the total force acting on the mass } m$$

$$= -\gamma\frac{d\xi}{dt} - k\xi$$

viscous + Hooke's law

or

$$m\frac{d^2\xi}{dt^2} + \gamma\frac{d\xi}{dt} + k\xi = 0$$

Substitution of $\xi(t) = e^{\alpha t}$ gives

$$m\alpha^2 + \gamma\alpha + k = 0$$

with

$$\alpha = -\frac{\gamma}{2m} \pm \frac{1}{2m}\sqrt{\gamma^2 - 4km}$$

$$= -\frac{\gamma}{2m} \pm \sqrt{\left[\frac{\gamma}{2m}\right]^2 - \frac{k}{m}}$$

The general solution is

$$\xi(t) = Ae^{-\gamma t/2m}\cos(\omega' t + \phi)$$

where

$$\omega' = \sqrt{\frac{k}{m} - \left[\frac{\gamma}{2m}\right]^2}$$

The motion is oscillatory with an amplitude that decreases with time if $k/m > (\gamma/2m)^2$. It is called damped oscillatory motion.

**2-9.** We substitute

$$y(x) = c_1y_1(x) + c_2y_2(x)$$

into

$$\frac{d^2y}{dx^2} + a_1(x)\frac{dy}{dx} + a_0(x)y(x) = 0$$

$$c_1\frac{d^2y_1}{dx^2} + c_2\frac{d^2y_2}{dx^2} + c_1a_1(x)\frac{dy_1}{dx} + c_2a_1(x)\frac{dy_2}{dx}$$
$$+ c_1a_0(x)y_1(x) + c_2a_0(x)y_2(x) \stackrel{?}{=} 0$$

$$c_1\left[\frac{d^2y_1}{dx^2} + a_1(x)\frac{dy_1}{dx} + a_0(x)y_1(x)\right]$$
$$+ c_2\left[\frac{d^2y_2}{dx^2} + a_1(x)\frac{dy_2}{dx} + a_0(x)y_2(x)\right] \stackrel{?}{=} 0$$

But $y_1(x)$ and $y_2(x)$ are each solutions to the differential equation, and so we see that

$$y(x) = c_1y_1(x) + c_2y_2(x)$$

is also a solution.

**2-10.** The general solution is

$$\psi(x) = A\sin\beta x + B\cos\beta x$$

where $\beta = (8\pi^2mE/h^2)^{1/2}$. The two boundary conditions give

$$\psi(0) = B = 0$$

$$\psi(a) = A\sin\beta a = 0$$

We require that $\sin\beta a = 0$, or that $\beta a = n\pi$, where $n = 1,2,\ldots$ Using the definition of $\beta$, we have

$$\beta^2 = \frac{8\pi^2mE}{h^2} = \frac{n^2\pi^2}{a^2}$$

or

$$E = \frac{n^2h^2}{8ma^2}$$

**2-11.** The form of the standing wave for the nth harmonic is $\sin(n\pi x/\ell)$. This function equals zero when x equals integral multiple values of $\ell/n$. We do not count the fixed points at $x = 0$ and $x = \ell$ as nodes, and so there are $n - 1$ nodes between $x = 0$ and $x = \ell$.

**2-12.** To prove that $y(x,t)$ is a wave of wavelength $\lambda$, replace $x$ by $x + \lambda$

$$y(x + \lambda,t) = A\sin\left[\frac{2\pi}{\lambda}\{(x + \lambda) - vt\}\right]$$

$$= A\sin\left[\frac{2\pi}{\lambda}(x-vt) + 2\pi\right]$$

$$= A\sin\left[\frac{2\pi}{\lambda}(x - vt)\right] = y(x,t)$$

To show that $y(x,t)$ oscillates with frequency $v/\lambda$, replace $t$ by $\lambda/v$, the period of the vibration.

$$y\left[x,t + \frac{\lambda}{v}\right] = A\sin\left[\frac{2\pi}{\lambda}\{x - v(t + \lambda/v)\}\right]$$

$$= A\sin\left[\frac{2\pi}{\lambda}(x - vt) + 2\pi\right]$$

$$= A\sin\left[\frac{2\pi}{\lambda}(x - vt)\right] = y(x,t)$$

To show that $y(x,t)$ is traveling to the right with a velocity $v$, look at any maximum of $y(x,t)$, say when the argument of the sine is equal to $\pi/2$. In this case

$$\frac{2\pi}{\lambda}(x - vt) = \frac{\pi}{2}$$

or

$$x = vt + \lambda/4$$

Note that $x$ increases linearly with $t$ with a speed $v$.

**2-13.** The form of the nth harmonic is

$$u_n(x,t) = A_n\cos\omega_n t\ \sin(n\pi x/\ell)$$

where $\omega_n = n\pi v/\ell$. Using the trigonometric identity

$$\sin\alpha\ \cos\beta = (1/2)\sin(\alpha + \beta) + (1/2)\sin(\alpha - \beta)$$

we write

$$u_n(x,t) = \frac{A_n}{2}\sin\left[\frac{n\pi x}{\ell} + \frac{n\pi vt}{\ell}\right] + \frac{A_n}{2}\sin\left[\frac{n\pi x}{\ell} - \frac{n\pi vt}{\ell}\right]$$

$$= \frac{A_n}{2}\sin\left[\frac{n\pi}{\ell}(x + vt)\right] + \frac{A_n}{2}\sin\left[\frac{n\pi}{\ell}(x - vt)\right]$$

which represent similar traveling waves traveling in opposite directions.

**2-14.**

$$\frac{\partial T}{\partial t} = \kappa^2\frac{\partial^2 T}{\partial x^2}$$

Let $T(x,t) = f(x)g(t)$ and get

$$\frac{1}{g(t)}\frac{dg}{dt} = \kappa^2\frac{1}{f(x)}\frac{d^2f}{dx^2} = \text{constant} = -\lambda^2$$

Then

$$g(t) = e^{-\lambda^2 t}$$

and

$$f(x) = A\cos\beta x + B\sin\beta x$$

where $\beta = \lambda/\kappa$. The end conditions of the bar give

$$f(0) = A = 0$$

$$f(\ell) = B\sin\beta\ell$$

We require that $\sin\beta\ell = 0$, or that $\beta\ell = n\pi$, where $n = 1, 2, \ldots$ Our solution for $T(x,t)$ so far is

$$\exp(-\lambda_n{}^2 t)\ B_n\sin\frac{n\pi x}{\ell}$$

and the complete solution is

$$T(x,t) = \sum_{n=1}^{\infty} B_n e^{-\lambda_n^2 t}\sin\frac{n\pi x}{\ell}$$

where $\lambda_n = \kappa n\pi/\ell$.

**2-15.** The nodes are given by the zeros of

$$(\sin n\pi x/a)(\sin m\pi y/b).$$

There are nodal lines when $x$ equals integral multiples of $a/n$ and when $y$ equals integral multiples of $b/m$. The first few normal modes look like those shown in Figure 2-6.

In one dimension (vibrating string), one gets nodal points. In two dimensions (vibrating membrane), one obtains nodal lines (as above). In three dimensions, one obtains nodal surfaces.

**2-16.** Let $\psi(x,y) = X(x)Y(y)$ and substitute this into the Schrödinger equation to obtain

$$\frac{1}{X(x)}\frac{d^2X}{dx^2} + \frac{1}{Y(y)}\frac{d^2Y}{dy^2} + \beta^2 = 0$$

where $\beta^2 = 8\pi^2mE/h^2$. Using the standard argument involving separation of variables, we write

$$\frac{d^2X}{dx^2} + p^2X(x) = 0 \qquad\qquad \frac{d^2Y}{dy^2} + q^2Y(y) = 0$$

where $p^2 + q^2 = \beta^2$. The solutions to these two differential equations are

$$X(x) = A\cos px + B\sin px$$

$$Y(y) = C\cos qy + D\sin qy$$

The boundary conditions are

$$X(0) = A = 0 \qquad\qquad Y(0) = C = 0$$

$$X(a) = B\sin pa = 0 \qquad\qquad Y(b) = D\sin qb = 0$$

from which we obtain $p = n_x\pi/a$ where $n_x = 1,2,\ldots$ and $q = n_y\pi/b$ where $n_y = 1,2,\ldots$ Using the fact that

$$\beta^2 = \frac{8\pi^2mE}{h^2} = p^2 + q^2 = \frac{n_x{}^2\pi^2}{a^2} + \frac{n_y{}^2\pi^2}{b^2}$$

we have

$$E_{n_xn_y} = \frac{n_x{}^2h^2}{8ma^2} + \frac{n_y{}^2h^2}{8mb^2}$$

**2-17.** This problem is similar to Problem 2-16. Let

$$\psi(x,y,z) = X(x)Y(y)Z(z)$$

and substitute this into

$$\frac{\partial^2\psi}{\partial x^2} + \frac{\partial^2\psi}{\partial y^2} + \frac{\partial^2\psi}{\partial z^2} = -\left[\frac{8\pi^2mE}{h^2}\right]\psi(x,y,z)$$

to obtain

$$Y(y)Z(z)\frac{\partial^2 X}{\partial x^2} + X(x)Z(z)\frac{\partial^2 Y}{\partial y^2} + X(x)Y(y)\frac{\partial^2 Z}{\partial z^2}$$

$$= - \left[\frac{8\pi^2 mE}{h^2}\right]XYZ$$

Dividing by $\psi = XYZ$ gives

$$\frac{1}{X}\frac{\partial^2 X}{\partial x^2} + \frac{1}{Y}\frac{\partial^2 Y}{\partial y^2} + \frac{1}{Z}\frac{\partial^2 Z}{\partial z^2} = -\frac{8\pi^2 mE}{h^2} \equiv -\beta^2$$

Since x,y,z are independent, then each term must equal a constant.

$$\frac{1}{X}\frac{\partial^2 X}{\partial x^2} = -p^2$$

$$\frac{1}{Y}\frac{\partial^2 Y}{\partial y^2} = -q^2$$

$$\frac{1}{Z}\frac{\partial^2 Z}{\partial z^2} = -s^2$$

where $p^2 + q^2 + s^2 = \beta^2$.

Thus we have three ordinary differential equations

$$\frac{d^2 X}{dx^2} + p^2 X(x) = 0$$

$$\frac{d^2 Y}{dy^2} + q^2 Y(y) = 0$$

$$\frac{d^2 Z}{dz^2} + s^2 Z(z) = 0$$

The solutions to these three differential equations are

$$X(x) = A\cos px + B\sin px$$

$$Y(y) = C\cos qy + D\sin qy$$

$$Z(z) = F\cos sz + G\sin sz$$

Imposing the boundary conditions

$$X(0) = X(a) = 0$$

$$Y(0) = Y(b) = 0$$

$$Z(0) = Z(c) = 0$$

leads to

$$A = 0, \; pa = n_x\pi$$

$$C = 0, \; qb = n_y\pi$$

$$F = 0, \; sc = n_z\pi$$

or

$$\psi(x,y,z) = X(x)Y(y)Z(z)$$

$$= H\sin\frac{n_x\pi x}{a}\,\sin\frac{n_y\pi y}{b}\,\sin\frac{n_z\pi z}{c}$$

Because $\beta^2 = p^2 + q^2 + s^2$ and

$$p^2 = \frac{n_x{}^2\pi^2}{a^2} \qquad q^2 = \frac{n_y{}^2\pi^2}{b^2} \qquad \text{and} \quad s^2 = \frac{n_z{}^2\pi^2}{c^2}$$

we have that

$$\frac{8\pi^2 mE}{h^2} = \frac{n_x{}^2\pi^2}{a^2} + \frac{n_y{}^2\pi^2}{b^2} + \frac{n_z{}^2\pi^2}{c^2}$$

or

$$E = \frac{h^2}{8m}\left[\frac{n_x{}^2}{a^2} + \frac{n_y{}^2}{b^2} + \frac{n_z{}^2}{c^2}\right]$$

**2-18.** The trajectory associated with Figure 2-9, $x(t) = (1/2)gt^2$, shows that a falling body is accelerated at a constant rate g. Now let's do the same problem using Figure 2-10.

Using Figure 2-10, we write

$$m\frac{d^2x}{dt^2} = -mg$$

This equation can be integrated twice to obtain

$$x(t) = -(1/2)gt^2 + c_1t + c_2$$

At $t = 0$, $x = x_0$ (see Figure 2-10), and so $c_2 = x_0$.

At $t = 0$,

$$v = \frac{dx}{dt} = -v_0$$

and so $c_1 = -v_0$ and

$$x(t) = -\frac{1}{2}gt^2 - v_0t + x_0$$

The value of x decreases with time in this case because x has opposite directions in Figures 2-9 and 2-10.

**2-19.** We let x be the height above the Earth's surface. Newton's equation is

$$m\frac{d^2x}{dt^2} = -mg$$

Integrating twice gives

$$x(t) = -\frac{1}{2}gt^2 + v_0t$$

because $x(0) = 0$ and $\dot{x}(0) = +v_0$. To find the time that it takes for the body to reach its maximum height, use

$$\frac{dx}{dt} = 0 = -gt_{max} + v_0$$

or

$$t_{max} = \frac{v_0}{g}$$

where $t_{max}$ is the time that it takes for x to reach $x_{max}$. The time that it takes the body to return is $2t_{max} = 2v_0/g$. The value of $x_{max}$, the maximum height attained, is given by

$$x_{max} = -\frac{1}{2}gt^2_{max} + v_0t_{max}$$

$$= -\frac{1}{2}g\left[\frac{v_0}{g}\right]^2 + v_0\left[\frac{v_0}{g}\right] = \frac{v_0^2}{2g}$$

**2-20.** From Figure 2-11, the component of mg in the direction of motion is $-mg\sin\theta$, the negative sign indicating that the force is in the opposite direction to the motion. Thus, Newton's equation is

$$m\ell\frac{d^2\theta}{dt^2} = -mg\sin\theta$$

If $\theta(t)$ is restricted to small angles, then $\sin\theta \approx \theta$ and

$$m\ell\frac{d^2\theta}{dt^2} = -mg\theta$$

or

$$\frac{d^2\theta}{dt^2} + \frac{g}{\ell}\theta(t) = 0$$

The general solution to this equation is

$$\theta(t) = A\sin(\omega t + \phi)$$

where $\omega = (g/\ell)^{1/2}$. Note that the natural frequency of the pendulum is independent of m.

**2-21.** Newton's equation is (see Problem 2-20)

$$m\ell\frac{d^2\theta}{dt^2} = -\lambda\frac{ds}{dt} - mg\sin\theta = -\lambda\ell\frac{d\theta}{dt} - mg\theta$$

for small angles. Thus we have

$$m\ell\frac{d^2\theta}{dt^2} + \lambda\ell\frac{d\theta}{dt} + mg\theta = 0$$

or

$$\frac{d^2\theta}{dt^2} + \frac{\lambda}{m}\frac{d\theta}{dt} + \frac{g}{\ell}\theta = 0$$

The general solution to this equation is found from

$$\alpha^2 + \frac{\lambda}{m}\alpha + \frac{g}{\ell} = 0$$

or

$$\alpha = -\frac{\lambda}{2m} \pm \frac{1}{2}\left[\frac{\lambda^2}{m^2} - \frac{4g}{\ell}\right]^{1/2}$$

and the general solution is

$$\theta(t) = e^{-\lambda t/2m}\left[c_1\exp\left[\left[\frac{\lambda^2}{4m^2} - \frac{g}{\ell}\right]^{1/2}t\right]\right.$$
$$\left. + c_2\exp\left[-\left[\frac{\lambda^2}{4m^2} - \frac{g}{\ell}\right]^{1/2}t\right]\right]$$

If

$$\frac{\lambda^2}{4m^2} > \frac{g}{\ell}$$

there is no oscillatory motion.

**2-22.** The quantities x(t) and y(t) are the distances along the arcs in Figure 2-12. In terms of angles (see Problem 2-20), we can write

$$m\ell\frac{d^2\theta_x}{dt^2} = -\ mg\theta_x - k(x - y)$$

where $x = \ell\theta_x$, $y = \ell\theta_y$ and we are assuming small displacements from the vertical. In terms of x and y we have

$$m\frac{d^2x}{dt^2} = -\ \frac{mg}{\ell}x - k(x - y)$$

and

$$m\frac{d^2y}{dt^2} = -\ \frac{mg}{\ell}y - k(y - x)$$

Notice that the force due to the connecting spring acts in opposite directions in the equations for x and y. If we substitute $x(t) = Ae^{i\omega t}$ and $y(t) = Be^{i\omega t}$ into these equations, then we find

$$\left[\omega^2 - \omega_0{}^2 - \frac{k}{m}\right]A = -\ \frac{k}{m}B$$

$$\left[\omega^2 - \omega_0{}^2 - \frac{k}{m}\right]B = -\ \frac{k}{m}A$$

where $\omega_0{}^2 = g/\ell$.

The determinant associated with the equations for A and B is

$$\begin{vmatrix} \omega^2 - \omega_0{}^2 - \frac{k}{m} & \frac{k}{m} \\ \frac{k}{m} & \omega^2 - \omega_0{}^2 - \frac{k}{m} \end{vmatrix} = \left[\omega^2 - \omega_0{}^2 - \frac{k}{m}\right]^2 - \left[\frac{k}{m}\right]^2$$

For this determinant to vanish, we must have

$$\omega_1^2 = \omega_0^2 \qquad \text{or} \qquad \omega_2^2 = \omega_0^2 + 2k/m$$

In the first case ($\omega_1 = \omega_0$), the two pendulums swing together in phase. In the second case, they swing against one another, 180° out of phase.

**2-23.** The two coupled differential equations from Problem 2-22 are

$$m\frac{d^2x}{dt^2} = -m\omega_0^2 x - k(x - y)$$

$$m\frac{d^2y}{dt^2} = -m\omega_0^2 y - k(y - x)$$

If we add these two equations, then we get

$$\frac{d^2}{dt^2}(x + y) + \omega_0^2(x + y) = 0$$

If we subtract them, we get

$$\frac{d^2}{dt^2}(x - y) + \omega_0^2(x - y) + \frac{2k}{m}(x - y) = 0$$

Introducing $\eta = x + y$ and $\xi = x - y$ gives

$$\frac{d^2\eta}{dt^2} + \omega_0^2\eta = 0$$

$$\frac{d^2\xi}{dt^2} + (\omega_0^2 + 2k/m)\xi = 0$$

These two equations are independent and are the same equations that are obtained in Problem 2-22.

Their integrated forms are

$$\eta(t) = A\sin(\omega_0 t + \phi)$$

$$\xi(t) = A'\sin(\omega t + \phi')$$

where $\omega^2 = \omega_0^2 + 2k/m$.

**2-24.** To prove that $f(x \pm vt)$ are solutions to the wave equation, let's investigate

$$\frac{\partial f}{\partial t} = \frac{\partial f}{\partial(x \pm vt)}\,\frac{\partial(x \pm vt)}{\partial t} = \pm v\frac{\partial f}{\partial(x \pm vt)}$$

Similarly

$$\frac{\partial^2 f}{\partial t^2} = v^2 \frac{\partial^2 f}{\partial (x \pm vt)^2}$$

and

$$\frac{\partial^2 f}{\partial x^2} = \frac{\partial^2 f}{\partial (x \pm vt)^2}$$

Note that

$$\frac{\partial^2 f}{\partial x^2} = \frac{1}{v^2} \frac{\partial^2 f}{\partial t^2}$$

becomes

$$\frac{\partial^2 f}{\partial (x \pm vt)^2} = \frac{1}{v^2} \left[ v^2 \frac{\partial^2 f}{\partial (x \pm vt)^2} \right] = \frac{\partial^2 f}{\partial (x \pm vt)^2}$$

Both $f(x + vt)$ and $f(x - vt)$ are waves traveling to the left and to the right, respectively.

Now take

$$\frac{\partial^2 u}{\partial x^2} = \frac{1}{v^2} \frac{\partial^2 u}{\partial t^2} = \frac{\partial^2 u}{\partial y^2}$$

where $y = vt$ and let

$$\eta = x + y \qquad \text{and} \qquad \xi = x - y$$

Then using the chain rule of partial differentiation

$$\frac{\partial u}{\partial x} = \frac{\partial u}{\partial \eta} \frac{\partial \eta}{\partial x} + \frac{\partial u}{\partial \xi} \frac{\partial \xi}{\partial x} = \frac{\partial u}{\partial \eta} + \frac{\partial u}{\partial \xi}$$

$$\frac{\partial^2 u}{\partial x^2} = \frac{\partial}{\partial x}\left[\frac{\partial u}{\partial \eta}\right] + \frac{\partial}{\partial x}\left[\frac{\partial u}{\partial \xi}\right] = \frac{\partial^2 u}{\partial \eta^2} + \frac{\partial^2 u}{\partial \eta \partial \xi} + \frac{\partial^2 u}{\partial \eta \partial \xi} + \frac{\partial^2 u}{\partial \xi^2}$$

and

$$\frac{\partial u}{\partial y} = \frac{\partial u}{\partial \eta} \frac{\partial \eta}{\partial y} + \frac{\partial u}{\partial \xi} \frac{\partial \xi}{\partial y} = \frac{\partial u}{\partial \eta} - \frac{\partial u}{\partial \xi}$$

$$\frac{\partial^2 u}{\partial y^2} = \frac{\partial}{\partial y}\left[\frac{\partial u}{\partial \eta}\right] - \frac{\partial}{\partial y}\left[\frac{\partial u}{\partial \xi}\right] = \frac{\partial^2 u}{\partial \eta^2} - \frac{\partial^2 u}{\partial \eta \partial \xi} - \frac{\partial^2 u}{\partial \eta \partial \xi} + \frac{\partial^2 u}{\partial \xi^2}$$

Substitute all this into the wave equation to obtain

$$\frac{\partial^2 u}{\partial \eta^2} + 2\frac{\partial^2 u}{\partial \eta \partial \xi} + \frac{\partial^2 u}{\partial \xi^2} = \frac{\partial^2 u}{\partial \eta^2} - 2\frac{\partial^2 u}{\partial \eta \partial \xi} + \frac{\partial^2 u}{\partial \xi^2}$$

or simply

$$\frac{\partial^2 u}{\partial\eta\partial\xi} = 0$$

A first integration of this equation gives

$$\frac{\partial u}{\partial\eta} = f(\eta)$$

and a second integration gives

$$u(\eta,\xi) = \int f(\eta)d\eta + \text{"const"} \equiv \phi(\eta) + \text{"const"},$$

but "const" is really an arbitrary function of $\xi$.

Thus,

$$u(\eta,\xi) = \phi(\eta) + \psi(\xi)$$

# CHAPTER THREE

## THE SCHRÖDINGER EQUATION AND SOME SIMPLE APPLICATIONS

3-1. a) $\hat{A}f = \text{SQRT}(x^4) = \pm x^2$

b)

$$\hat{A}f = \left[\frac{d^3}{dx^3} + x^3\right]e^{-\alpha x} = (x^3 - \alpha^3)e^{-\alpha x}$$

c)

$$\hat{A}f = \int_0^1 dx\,(x^3 - 2x + 3) = \frac{1}{4} - \frac{2}{2} + 3 = \frac{9}{4}$$

d)

$$\hat{A}f = \left[\frac{\partial^2}{\partial x^2} + \frac{\partial^2}{\partial y^2} + \frac{\partial^2}{\partial z^2}\right]x^3y^2z^4$$

$$= 6xy^2z^4 + 2x^3z^4 + 12x^3y^2z^2$$

3-2.

a) $\hat{A}[c_1f_1(x) + c_2f_2(x)]$

$= \text{SQR}[c_1f_1(x) + c_2f_2(x)]$

$= [c_1f_1(x) + c_2f_2(x)]^2$

$= c_1{}^2f_1{}^2(x) + c_2{}^2f_2{}^2(x) + 2c_1c_2f_1(x)f_2(x)$

$\neq c_1[f_1(x)]^2 + c_2[f_2(x)]^2$

Therefore, $\hat{A}$ is nonlinear.

b) $\hat{A}[c_1f_1(x) + c_2f_2(x)]$

$= [c_1f_1(x) + c_2f_2(x)]^*$

$= c_1{}^*f_1{}^*(x) + c_2{}^*f_2{}^*(x)$

$= c_1{}^*\hat{A}f_1(x) + c_2{}^*\hat{A}f_2(x)$

$\neq c_1\hat{A}f_1(x) + c_2\hat{A}f_2(x)$

Therefore, $\hat{A}$ is nonlinear.

c) $\hat{A}[c_1f_1(x) + c_2f_2(x)]$

$= (0)[c_1f_1(x) + c_2f_2(x)]$

$= c_1(0)f_1(x) + c_2(0)f_2(x)$

$= 0$

Therefore, $\hat{A}$ is linear.

d) $\hat{A}[c_1f_1(x) + c_2f_2(x)]$

$$= [c_1f_1(x) + c_2f_2(x)]^{-1}$$

$$= \frac{1}{c_1f_1(x) + c_2f_2(x)}$$

$$\neq c_1[f_1(x)]^{-1} + c_2[f_2(x)]^{-1}$$

Therefore, $\hat{A}$ is nonlinear.

e) $\hat{A}[c_1f_1(x) + c_2f_2(x)] = c_1f_1(0) + c_2f_2(0)$

$$= c_1\hat{A}f_1(x) + c_2\hat{A}f_2(x)$$

Therefore, $\hat{A}$ is linear.

f) $\hat{A}[c_1f_1(x) + c_2f_2(x)]$

$$= \ln[c_1f_1(x) + c_2f_2(x)]$$

$$\neq c_1\ln f_1(x) + c_2\ln f_2(x)$$

Therefore, $\hat{A}$ is nonlinear.

**3-3.**

a)

$$\hat{A}f = \frac{d^2}{dx^2}\cos\omega x = -\omega^2\cos\omega x$$

b)

$$\hat{A}f = \frac{d}{dt}e^{i\omega t} = i\omega e^{i\omega t}$$

c)

$$\hat{A}f = \left[\frac{d^2}{dx^2} + 2\frac{d}{dx} + 3\right]e^{\alpha x} = (\alpha^2 + 2\alpha + 3)e^{\alpha x}$$

d)

$$\hat{A}f = \frac{\partial}{\partial y}x^2e^{6y} = 6x^2e^{6y}$$

**3-4.**

$$\nabla^2(\cos ax)(\cos by)(\cos cz)$$

$$= \left[\frac{\partial^2}{\partial x^2} + \frac{\partial^2}{\partial y^2} + \frac{\partial^2}{\partial z^2}\right](\cos ax)(\cos by)(\cos cz)$$

$$= (-a^2 - b^2 - c^2)(\cos ax)(\cos by)(\cos cz)$$

3-5.

a)

$$\hat{A}^2 f = \hat{A}(\hat{A}f) = \frac{d^2}{dx^2}\left[\frac{d^2 f}{dx^2}\right] = \frac{d^4 f}{dx^4}$$

and so

$$\hat{A}^2 = \frac{d^4}{dx^4}$$

b)

$$\hat{A}^2 f = \left[\frac{d}{dx} + x\right]\left[\frac{df}{dx} + xf\right] = \frac{d^2 f}{dx^2} + 2x\frac{df}{dx} + f + x^2 f$$

Therefore,

$$\hat{A}^2 = \frac{d^2}{dx^2} + 2x\frac{d}{dx} + (x^2 + 1)$$

c)

$$\hat{A}^2 f = \left[\frac{d^2}{dx^2} - 2x\frac{d}{dx} + 1\right]\left[\frac{d^2 f}{dx^2} - 2x\frac{df}{dx} + f\right]$$

$$= \frac{d^4 f}{dx^4} - 4x\frac{d^3 f}{dx^3} + (4x^2 - 2)\frac{d^2 f}{dx^2} + f$$

Therefore,

$$\hat{A}^2 = \frac{d^4}{dx^4} - 4x\frac{d^3}{dx^3} + (4x^2 - 2)\frac{d^2}{dx^2} + 1$$

3-6.

a)

$$\hat{A}\hat{B}(f(x)) = \hat{A}(\hat{B}f(x)) = \frac{d}{dx}\left[\frac{d^2}{dx^2} + 2\frac{d}{dx}\right]f(x)$$

$$= \left[\frac{d^3}{dx^3} + 2\frac{d^2}{dx^2}\right]f(x)$$

and so

$$\hat{A}\hat{B} = \frac{d^3}{dx^3} + 2\frac{d^2}{dx^2}$$

$$\hat{B}\hat{A}f(x) = \hat{B}(\hat{A}f(x)) = \left[\frac{d^2}{dx^2} + 2\frac{d}{dx}\right]\left[\frac{d}{dx}f(x)\right]$$

$$= \left[\frac{d^3}{dx^3} + 2\frac{d^2}{dx^2}\right]f(x)$$

and so

$$\hat{B}\hat{A} = \frac{d^3}{dx^3} + 2\frac{d^2}{dx^2}$$

Since $\hat{A}\hat{B} = \hat{B}\hat{A}$, the operators commute.

b)

$$\hat{A}\hat{B}f = x\frac{df}{dx}$$

$$\hat{B}\hat{A}f = \frac{d}{dx}(xf) = x\frac{df}{dx} + f$$

Therefore,

$$\hat{A}\hat{B} = x\frac{d}{dx}$$

and

$$\hat{B}\hat{A} = 1 + x\frac{d}{dx}$$

Therefore $\hat{A}$ and $\hat{B}$ do not commute.

c) $\hat{A}\hat{B}f(x) = \text{SQR[SQRT}f(x)] = \text{SQR}[\pm f(x)^{1/2}]$

$$= f(x)$$

and so

$$\hat{A}\hat{B} = 1$$

$$\hat{B}\hat{A}f(x) = \text{SQRT[SQR}f(x)] = \text{SQRT}[f(x)]^2$$

$$= \pm f(x)$$

and so

$$\hat{B}\hat{A} = \pm 1$$

Therefore

$$\hat{A}\hat{B} \neq \hat{B}\hat{A}$$

and SQR and SQRT do not commute.

d)

$$\hat{A}\hat{B} = \left[\frac{\partial}{\partial x}\ \frac{\partial}{\partial y}\right]f = \frac{\partial^2 f}{\partial x\,\partial y}$$

$$\hat{B}\hat{A} = \left[\frac{\partial}{\partial y}\ \frac{\partial}{\partial x}\right]f = \frac{\partial^2 f}{\partial y\,\partial x}$$

For any well-behaved f(x,y)

$$\hat{A}\hat{B} = \hat{B}\hat{A}$$

and

$\frac{\partial}{\partial x}$ and $\frac{\partial}{\partial y}$ commute.

3-7.

$$(\hat{P} + \hat{Q})(\hat{P} - \hat{Q}) = \hat{P}\hat{P} + \hat{P}(-\hat{Q}) + \hat{Q}\hat{P} + \hat{Q}(-\hat{Q})$$
$$= \hat{P}^2 - \hat{Q}^2 + \hat{Q}\hat{P} - \hat{P}\hat{Q}$$

Only when $\hat{P}$ and $\hat{Q}$ commute do we obtain the same result as ordinary algebra.

3-8. We start with

$$E = \frac{h^2n^2}{8ma^2} = \frac{p^2}{2m}$$

Solve for p to obtain

$$p = \frac{nh}{2a}$$

The deBroglie wave lengths associated with these momenta are

$$\lambda = \frac{h}{p} = \frac{2a}{n}$$

Solving for a, we obtain

$$a = n\left[\frac{\lambda}{2}\right]$$

which shows that an integral number of half wavelengths fit into the box.

3-9. The skeletal formula of hexatriene is

$$\rangle C{=}C{-}C{=}C{-}C{=}C\langle$$

and it has $6\pi$ electrons.

The length is taken to be

| | | |
|---|---|---|
| 3 double bonds | → 3 × 1.35Å | = 4.05Å |
| 2 single bonds | → 2 × 1.54Å | = 3.08Å |
| 2 half single bonds | → 2 × 1/2 × 1.54Å | = 1.54Å |
| | | 8.67Å = length |

The six $\pi$ electrons fill the first three levels, so a single electronic transition will occur from the $n = 3$ level to the $n = 4$ level.

The energy difference is given by

$$\Delta E = \frac{h^2}{8ma^2}\left[4^2 - 3^2\right]$$

$$= \frac{(6.626 \times 10^{-34}\ \mathrm{J\cdot s})^2(7)}{8(9.11 \times 10^{-31}\ \mathrm{kg})(8.67 \times 10^{-10}\ \mathrm{m})^2}$$

$$= 5.61 \times 10^{-19}\ \mathrm{J}$$

This energy corresponds to

$$(5.61 \times 10^{-19}\ \mathrm{J})\left[\frac{5.035 \times 10^{22}\ \mathrm{cm^{-1}}}{\mathrm{J}}\right]$$

$$= 2.82 \times 10^4\ \mathrm{cm^{-1}}$$

**3-10.** Because $\hat{H}$ is a linear operator,

$$\hat{H}[c\psi(x)] = c\hat{H}\psi(x) = cE\psi(x) = E[c\psi(x)]$$

**3-11.**

$$\langle x\rangle = \sum_j x_j f(x_j)$$

$$= (-6)(0.05) - 2(0.15) + 0(0.50) + 1(0.10) + 3(0.05) + 4(0.10) + 5(0.05)$$

$$= 0.30$$

$$\langle x^2\rangle = \sum_j x_j^2 f(x_j)$$

$$= 36(0.05) + 4(0.15) + 0(0.50) + 1(0.10) + 9(0.05) + 16(0.10) + 25(0.05)$$

$$= 5.80$$

$$\sigma^2 = \langle x^2\rangle - \langle x\rangle^2 = 5.80 - (0.30)^2 = 5.71$$

**3-12.** To show that $f_n$ is normalized,

$$\sum_{n=0}^{\infty} f_n = \sum_{n=0}^{\infty} \frac{\lambda^n}{n!}e^{-\lambda} = e^{-\lambda}\sum_{n=0}^{\infty}\frac{\lambda^n}{n!} = e^{-\lambda}e^{\lambda} = 1$$

To find $\langle n \rangle$ and $\langle n^2 \rangle$ it is best to use the relations

$$e^{\lambda} = \frac{d}{d\lambda} e^{\lambda} = \frac{d}{d\lambda} \sum_{n=0}^{\infty} \frac{\lambda^n}{n!} = \sum_{n=0}^{\infty} \frac{n\lambda^{n-1}}{n!}$$

and

$$e^{\lambda} = \frac{d^2}{d\lambda^2} e^{\lambda} = \frac{d^2}{d\lambda^2} \sum_{n=0}^{\infty} \frac{\lambda^n}{n!} = \sum_{n=0}^{\infty} \frac{n(n-1)\lambda^{n-2}}{n!}$$

So

$$\langle n \rangle = \sum_{n=0}^{\infty} n f_n = \sum_{n=0}^{\infty} \frac{n\lambda^n}{n!} e^{-\lambda} = \lambda e^{-\lambda} \sum_{n=0}^{\infty} \frac{n\lambda^{n-1}}{n!}$$

$$= (\lambda e^{-\lambda}) e^{\lambda} = \lambda$$

$$\langle n^2 \rangle = \sum_{n=0}^{\infty} n^2 f_n = \sum_{n=0}^{\infty} \frac{n^2\lambda^n}{n!} e^{-\lambda} = \lambda^2 e^{-\lambda} \sum_{n=0}^{\infty} \frac{n^2\lambda^{n-2}}{n!}$$

To use the second-derivative expression that we derived above, subtract and add

$$\sum_{n=0}^{\infty} \frac{n\lambda^{n-2}}{n!} = \frac{1}{\lambda} \sum_{n=0}^{\infty} \frac{n\lambda^{n-1}}{n!} = \frac{e^{\lambda}}{\lambda}$$

to get

$$\langle n^2 \rangle = \lambda^2 e^{-\lambda} \left[ \sum_{n=0}^{\infty} \frac{(n^2 - n)\lambda^{n-2}}{n!} + \sum_{n=0}^{\infty} \frac{n\lambda^{n-2}}{n!} \right]$$

$$= \lambda^2 e^{-\lambda} \left[ e^{\lambda} + \frac{e^{\lambda}}{\lambda} \right] = \lambda^2 + \lambda$$

$$\sigma^2 = \langle n^2 \rangle - \langle n \rangle^2 = (\lambda^2 + \lambda) - \lambda^2 = \lambda > 0$$

**3-13.** $p(x)dx = ce^{-\lambda x}dx \qquad 0 \le x < \infty$

To find c, use the fact that p(x) is normalized

$$\int_0^{\infty} p(x)dx = 1 = c\int_0^{\infty} e^{-\lambda x}dx = \frac{c}{\lambda}$$

Therefore, $c = \lambda$.

$$\langle x \rangle = \int_0^{\infty} x\, p(x)dx = \lambda \int_0^{\infty} xe^{-\lambda x}dx = \frac{1}{\lambda}$$

$$\langle x^2\rangle = \int_0^\infty x^2 p(x)\,dx = \lambda\int_0^\infty x^2 e^{-\lambda x}dx = \frac{\lambda\cdot 2!}{\lambda^3} = \frac{2}{\lambda^2}$$

$$\sigma^2 = \langle x^2\rangle - \langle x\rangle^2 = \frac{1}{\lambda^2}$$

and

$$\text{prob}(x \geq a) = \int_a^\infty p(x)\,dx = \lambda\int_a^\infty e^{-\lambda x}dx = e^{-\lambda a}$$

**3-14.** The odd moments of a Gaussian distribution are given by

$$\langle x^{2n+1}\rangle = (2\pi\sigma^2)^{-1/2}\int_{-\infty}^{\infty} x^{2n+1}\exp(-x^2/2\sigma^2)\,dx$$

where n is an integer.

The integrands are odd functions of x and the limits are symmetric; therefore $\langle x^{2n+1}\rangle = 0$. For $\langle x^4\rangle$, we have

$$\langle x^4\rangle = (2\pi\sigma^2)^{-1/2}\int_{-\infty}^{\infty} x^4\exp(-x^2/2\sigma^2)\,dx$$

From almost any table of integrals, we find

$$\int_0^\infty x^{2n}e^{-ax^2}dx = \frac{1\cdot 3\cdots(2n-1)}{2^{n+1}a^n}\left[\frac{\pi}{a}\right]^{1/2} \qquad n \geq 1$$

For n = 2, we have

$$\int_0^\infty x^4 e^{-ax^2}dx = \frac{1\cdot 3}{8a^2}\left[\frac{\pi}{a}\right]^{1/2}$$

Because the integrand is an even function of x

$$\int_0^\infty x^4 e^{-ax^2}dx = \frac{1}{2}\int_{-\infty}^{\infty} x^4 e^{-ax^2}dx$$

Therefore,

$$\langle x^4\rangle = (2\pi\sigma^2)^{-1/2}\int_{-\infty}^{\infty} x^4 e^{-x^2/2\sigma^2}dx$$

$$= \frac{2}{(2\pi\sigma^2)^{1/2}}\,\frac{3}{8}(2\sigma^2)^2(2\sigma^2\pi)^{1/2} = 3\sigma^4$$

3-15. Start with

$$I(\beta) = \int_0^a e^{\beta x}\sin^2 \frac{n\pi x}{a}dx$$

Set $\beta = 0$ and write

$$I(0) = \int_0^a \sin^2 \frac{n\pi x}{a}dx$$

Differentiate $I(\beta)$ with respect to $\beta$

$$\frac{dI}{d\beta} = \int_0^a xe^{\beta x}\sin^2 \frac{n\pi x}{a}dx$$

and set $\beta = 0$ to get

$$I'(0) = \int_0^a x \sin^2 \frac{n\pi x}{a}dx$$

Lastly, we write

$$\frac{d^2I}{d\beta^2} = \int_0^a x^2e^{\beta x}\sin^2 \frac{n\pi x}{a}dx$$

from which we get

$$I''(0) = \int_0^a x^2\sin^2 \frac{n\pi x}{a}dx$$

The value of $I(\beta)$ is obtained from a table of integrals

$$\int e^{\alpha x}\sin^2bx\ dx$$

$$= \frac{e^{\alpha x}}{2\alpha} - \frac{e^{\alpha x}}{\alpha^2 + 4b^2}\left[\frac{\alpha}{2}\cos2bx + b\sin2bx\right]$$

Using this result,

$$I(\beta) = \int_0^a e^{\beta x}\sin^2\frac{n\pi x}{a}dx$$

$$= \left|\frac{e^{\beta x}}{2\beta} - \frac{e^{\beta x}}{\beta^2 + (4n^2\pi^2/a^2)} \times \left[\frac{\beta}{2}\cos\frac{2\pi nx}{a} + \frac{n\pi}{a}\sin\frac{2\pi nx}{a}\right]\right|_0^a$$

$$= \frac{e^{\beta a} - 1}{2\beta} - \frac{(\beta/2)(e^{\beta a} - 1)}{\beta^2 + (4n^2\pi^2/a^2)}$$

To evaluate $I(0)$, $I'(0)$ and $I''(0)$, it is easiest to use the relation (Maclaurin series)

$$I(\beta) = I(0) + \beta I'(0) + (\beta^2/2)I''(0) + \cdots$$

By using the series expansion of $e^{\beta a}$, we obtain

$$I(\beta) = \frac{a}{2} + \frac{a^2}{4}\beta + \frac{a^3}{12}\beta^2 + \cdots - \frac{(\beta^2 a/2) + (\beta^3 a^2/4) + \cdots}{(\beta^2 + 4\pi^2 n^2/a^2)}$$

$$= \frac{a}{2} + \frac{a^2}{4}\beta + \frac{a^3}{12}\beta^2 + \cdots - \frac{a^2}{4\pi^2 n^2}\left[\frac{a}{2}\beta^2 + \frac{a^2}{4}\beta^3 + \cdots\right]\left[1 - \frac{a^2\beta^2}{4\pi^2 n^2} + \cdots\right]$$

$$= \frac{a}{2} + \frac{a^2}{4}\beta + \left[\frac{a^3}{6} - \frac{a^3}{4\pi^2 n^2}\right]\frac{\beta^2}{2} + \cdots$$

Therefore,

$$I(0) = a/2$$

$$I'(0) = (a^2/4)$$

$$I''(0) = \frac{a^3}{6} - \frac{a^3}{4\pi^2 n^2} = \left[\frac{a}{2\pi n}\right]^3\left[\frac{4\pi^3 n^3}{3} - 2\pi n\right]$$

**3-16.** We must evaluate

$$\langle x \rangle = \int_0^a \psi_n^*(x)\, x\, \psi_n(x)\,dx$$

$$= \frac{2}{a}\int_0^a x \sin^2 \frac{n\pi x}{a}\,dx$$

$$= \frac{2}{a}\left|\frac{x^2}{4} - \frac{x\sin(2\pi n x/a)}{4n\pi/a} - \frac{\cos(2\pi n x/a)}{8n^2\pi^2/a^2}\right|_0^a$$

$$= \frac{2}{a}\left[\frac{a^2}{4} - \frac{a^2}{8n^2\pi^2} + \frac{a^2}{8n^2\pi^2}\right] = \frac{a}{2}$$

The average value of x is midway in the box.

3-17. From Eq. 3-47,

$$\sigma_x = \frac{a}{2\pi n}\left[\frac{\pi^2 n^2}{3} - 2\right]^{1/2} = a\left[\frac{\pi^2 n^2 - 6}{12\pi^2 n^2}\right]^{1/2}$$

The quantity that multiplies a is less than unity for all values of n.

3-18.

$$\int_0^a \sin\frac{n\pi x}{a}\cos\frac{n\pi x}{a}dx = \frac{1}{2}\int_0^a \sin\frac{2\pi nx}{a}dx$$

$$= \frac{1}{2}\left|_0^a - \frac{a\cos(2\pi nx/a)}{2\pi n}\right| = 0$$

3-19. Eq. 3-15 for $\hat{K}$ is

$$\hat{K} = -\frac{\hbar}{2m}\frac{d^2}{dx^2} = \frac{1}{2m}\hat{P}_x{}^2$$

Thus

$$\langle K\rangle = \frac{1}{2m}\langle p^2\rangle$$

Using Eq. 3-60,

$$\langle K\rangle = \frac{1}{2m}\left[\frac{n^2\pi^2\hbar^2}{a^2}\right] = \frac{n^2h^2}{8ma^2}$$

Note that $\langle K\rangle = E$ for a particle in a box.

3-20.

$$\int_0^a e^{\pm i2\pi nx/a}dx = \int_0^a \cos\frac{2\pi nx}{a}dx \pm i\int_0^a \sin\frac{2\pi nx}{a}dx$$

Both integrals here go over complete cycles of the cosine and sine, and hence vanish.

3-21. Start with Eq. 3-61.

$$\sigma_p = n\pi\hbar/a$$

If $a = 10^{-14}$m then

$$\sigma_p \approx 3 \times 10^{-20}\ \text{kg}\cdot\text{m}\cdot\text{s}^{-1}$$

Using the relation

$$E = \frac{\sigma_p^{\,2}}{2m}$$

as an estimate of the uncertainy in E,

$$E = \frac{(3 \times 10^{-20}\ \text{kg·m·s}^{-1})^2}{2 \times 9.11 \times 10^{-31}\ \text{kg}} \approx 5 \times 10^{-10}\ \text{J}$$

$$= 3\times10^{9}\ \text{eV} = 3000\ \text{MeV}$$

**3-22.** We wish to show that

$$I_{nm} = \int_0^a \psi_n(x)\psi_m(x)dx = \frac{2}{a}\int_0^a \sin\frac{n\pi x}{a}\sin\frac{m\pi x}{a}dx = \delta_{nm}$$

Use the trigonometric identity

$$\sin\frac{n\pi x}{a}\sin\frac{m\pi x}{a} = \frac{1}{2}\cos\frac{(n-m)\pi x}{a} - \frac{1}{2}\cos\frac{(m+n)\pi x}{a}$$

to write

$$I_{nm} = \frac{1}{a}\int_0^a \cos\frac{(n-m)\pi x}{a}dx - \frac{1}{a}\int_0^a \cos\frac{(n+m)\pi x}{a}dx$$

$$= \frac{1}{a}\left|_0^a \frac{a\sin[(n-m)\pi x/a]}{(n-m)\pi}\right|$$

$$- \frac{1}{a}\left|_0^a \frac{a\sin[(n+m)\pi x/a]}{(n+m)\pi}\right| \qquad n \neq m$$

$$= \frac{1}{a}(0) - \frac{1}{a}(0) = 0 \qquad n \neq m$$

where we have used the fact that $\sin N\pi = 0$ for integral values of N.

Note that if n = m, we have

$$I_{nn} = \frac{1}{a}\int_0^a dx - \frac{1}{a}\int_0^a \cos\frac{2\pi nx}{a}dx = 1$$

**3-23.** The orthonormality condition is

$$\int_0^a \psi_n^*(x)\psi_m(x)dx = \delta_{nm}$$

For $n = m$, we have

$$\frac{1}{a}\int_0^a e^{-in\pi x/a}\, e^{in\pi x/a} = \frac{1}{a}\int_0^a dx = 1$$

For $n \neq m$,

$$\frac{1}{a}\int_0^a e^{i\pi(m-n)x/a}dx$$

$$= \frac{1}{a}\int_0^a \left[\cos\frac{\pi(m-n)x}{a} + i\,\sin\frac{\pi(m-n)x}{a}\right]dx = 0$$

(see Problem 3-22)

**3-24.**

$$\int_0^a \sin\frac{n\pi x}{a}\sin\frac{m\pi x}{a}dx$$

$$= \frac{1}{(2i)^2}\int_0^a \left[(e^{in\pi x/a} - e^{-in\pi x/a}) \times (e^{im\pi x/a} - e^{-im\pi x/a})\right]dx$$

$$= -\frac{1}{4}\int_0^a \left[e^{i(n+m)\pi x/a} - e^{i(m-n)\pi x/a} - e^{i(n-m)\pi x/a} + e^{-i(n+m)\pi x/a}\right]dx$$

$= 0$ if $m \neq n$ (see Problem 3-23)

$= 1/2$ if $n = m$

Similarly

$$\int_0^a \cos\frac{n\pi x}{a}\cos\frac{m\pi x}{a}dx$$

$$= \frac{1}{4}\int_0^a (e^{in\pi x/a} + e^{-in\pi x/a})(e^{im\pi x/a} + e^{-im\pi x/a})dx$$

$$= \frac{1}{4}\int_0^a \left[e^{i(n+m)\pi x/a} + e^{i(m-n)\pi x/a} + e^{i(n-m)\pi x/a} + e^{-i(n+m)\pi x/a}\right]dx$$

$= 0$ if $n \neq m$ (see Problem 3-23)

$= 1/2$ if $n = m$

**3-25.** For the case m = n, we have

$$\int_0^{2\pi} \phi_n^*(\theta)\phi_n(\theta)d\theta$$

$$= \int_0^{2\pi} [(2\pi)^{-1/2}e^{-in\theta}][(2\pi)^{-1/2}e^{in\theta}]d\theta$$

$$= \frac{1}{2\pi}\int_0^{2\pi} d\theta = 1$$

For the case $m \neq n$, we have

$$\int_0^{2\pi} \phi_m^*(\theta)\phi_n(\theta)d\theta = \frac{1}{2\pi}\int_0^{2\pi} e^{i(n-m)\theta}d\theta = 0$$

**3-26.** We must solve

$$\frac{d^2\psi}{d\theta^2} + \frac{2IE}{\hbar^2}\psi = 0 \qquad\qquad 0 \leq \theta \leq 2\pi$$

The solutions to this equation are

$$\psi(\theta) = \exp(\pm\, i(2IE)^{1/2}\theta/\hbar) \equiv e^{\pm in\theta}$$

where $n = (2IE)^{1/2}/\hbar$

By imposing a continuity condition on $\psi(\theta)$, we obtain

$$\psi(\theta) = \psi(\theta + 2\pi)$$

or

$$e^{in\theta} = e^{in(\theta+2\pi)}$$

or

$$1 = e^{i2\pi n}$$

This condition implies that n must be 0, ±1, ±2, ...

Thus we have

$$E = \frac{n^2\hbar^2}{2I} \qquad\qquad n = 0, \pm 1, \pm 2, \ldots$$

There are six π electrons in benzene. We place two of them into the n = 0 level, two into the n = 1 level and two into the n = −1 level. The first electronic transition would be a 1→2 transition, and the frequency associated with this transition is

$$\bar{\nu} = \frac{h}{8\pi^2 cI}(2^2 - 1^2)$$

**3-27.** We must solve

$$\frac{d^2\psi}{dx^2} + \frac{2mE}{\hbar^2}\psi = 0 \qquad -a \le x \le a$$

where

$$\psi(a) = \psi(-a) = 0$$

The general solution is

$$\psi(x) = A\cos kx + B\sin kx$$

where $k = (2mE/\hbar^2)^{1/2}$. The boundary conditions give

$$\psi(a) = 0 = A\cos ka + B\sin ka$$

$$\psi(-a) = 0 = A\cos ka - B\sin ka$$

To obtain a non-trivial solution to these equations, we must have

$$\begin{vmatrix} \cos ka & \sin ka \\ \cos ka & -\sin ka \end{vmatrix} = -2\cos ka \sin ka = 0$$

We can write this condition as

$$\sin 2ka = 0$$

which gives

$$2ka = n\pi \qquad n = 1,2,\ldots$$

or

$$k = n\pi/2a \qquad n = 1,2,\ldots$$

If we substitute this result back into the two simultaneous equations for A and B, then we obtain

$$A\cos\frac{n\pi}{2} + B\sin\frac{n\pi}{2} = 0$$

$$A\cos\frac{n\pi}{2} - B\sin\frac{n\pi}{2} = 0$$

For odd values of n, we must have B = 0, while for even values of n, we must have A = 0.

Thus, our solutions are

$$\psi_n(x) = A\cos\frac{n\pi x}{2a} \qquad n \text{ odd}$$

$$= B\sin\frac{n\pi x}{2a} \qquad n \text{ even}$$

In either case, the normalization constant is $a^{-1/2}$.

The energies are given by

$$k = \frac{n\pi}{2a} = \left[\frac{2mE}{\hbar^2}\right]^{1/2}$$

or

$$E = \frac{n^2h^2}{32ma^2} = \frac{n^2h^2}{8m(2a)^2}$$

Certainly no experimentally observable properties will depend upon where <u>we</u> put the origin.

**3-28.** For n odd,

$$\psi_n(-x) = a^{-1/2}\cos\left[-\frac{n\pi x}{2a}\right] = a^{-1/2}\cos\frac{n\pi x}{2a} = \psi_n(x)$$

and for n even

$$\psi_n(-x) = a^{-1/2}\sin\left[-\frac{n\pi x}{2a}\right] = -a^{-1/2}\sin\frac{n\pi x}{2a} = -\psi_n(x)$$

We now show that $\hat{H}(x)$ is an even function of x.

Because

$$\hat{H}(x) = -\frac{\hbar^2}{2m}\frac{d^2}{dx^2}$$

then

$$\hat{H}(-x) = -\frac{\hbar^2}{2m}\frac{d^2}{d(-x)^2} = -\frac{\hbar^2}{2m}\frac{d^2}{dx^2} = \hat{H}(x)$$

and $\psi_n(x)$ and $\psi_n(-x)$ must differ by only a multiplicative constant for any nondegenerate state, or

$$\psi_n(-x) = c\psi_n(x)$$

By applying the inversion or reflection operation again, we obtain

$$\psi_n(x) = c\psi_n(-x) = c^2\psi_n(x)$$

or that $c = \pm 1$. Thus the wave functions are even or odd functions of x because of the symmetry of $\hat{A}(x)$.

**3-29.** As $n \to \infty$,

$$\sigma^2 = \frac{a^2}{12}\left[1 - \frac{6}{\pi^2 n^2}\right] \to \frac{a^2}{12}$$

The strictly classical value is given by the equations

$$\langle x \rangle = \int_0^a x\ p(x)dx = \frac{1}{a}\int_0^a x\ dx = \frac{a}{2}$$

$$\langle x^2 \rangle = \int_0^a x^2 p(x)dx = \frac{1}{a}\int_0^a x^2 dx = \frac{a^2}{3}$$

$$\sigma^2 = \langle x^2 \rangle - \langle x \rangle^2 = \frac{a^2}{12}$$

which agrees with the quantum mechanical result in the limit of large quantum number n. The agreement between classical mechanics and quantum mechanics in the limit of large quantum numbers is a general result.

**3-30.** Start with (see Figure 3-7)

$$(ds)^2 = (dx)^2 + (du)^2$$

Now divide by $(dx)^2$ to obtain

$$\left[\frac{ds}{dx}\right]^2 = 1 + \left[\frac{du}{dx}\right]^2$$

or

$$\frac{ds}{dx} = \left[1 + \left[\frac{du}{dx}\right]^2\right]^{1/2}$$

The potential energy is given by

$$U_n = \int_0^{\ell} T(ds - dx) = \int_0^{\ell} T\left[\frac{ds}{dx} - 1\right]dx$$

Substituting for $\frac{ds}{dx}$, we obtain

$$U_n = \int_0^\ell T\left[\left[1 + \left[\frac{\partial u}{\partial x}\right]^2\right]^{1/2} - 1\right]dx$$

Now we use the fact that $(1 + x)^{1/2} \approx (1 + (x/2))$ for small x

$$U_n = \int_0^\ell T\left[\left[1 + \frac{1}{2}\left[\frac{\partial u}{\partial x}\right]^2 + \cdots\right] - 1\right]dx$$

$$= \frac{1}{2}T\int_0^\ell \left[\frac{\partial u}{\partial x}\right]^2 dx$$

The kinetic energy is given by

$$K_n = \frac{\rho}{2}\int_0^\ell \left[\frac{\partial u}{\partial t}\right]^2 dx$$

$$= \frac{\rho}{2}\int_0^\ell D_n{}^2\omega_n{}^2\sin^2(\omega_n t + \phi_n)\sin^2\left[\frac{n\pi x}{\ell}\right]dx$$

$$= \frac{\pi^2 v^2 n^2 \rho}{2\ell^2}D_n{}^2\sin^2(\omega_n t + \phi_n)\left[\int_0^\ell \sin^2\left[\frac{n\pi x}{\ell}\right]dx\right]$$

$$= \frac{\pi^2 v^2 n^2 \rho}{4\ell}D_n{}^2\sin^2(\omega_n t + \phi_n)$$

Similarly we have

$$U_n = \frac{T}{2}\int_0^\ell \left[\frac{\partial u}{\partial x}\right]^2 dx$$

$$= D_n^2\frac{T}{2}\int_0^\ell \left[\frac{n\pi}{\ell}\cos(\omega_n t + \phi_n)\cos\frac{n\pi x}{\ell}\right]^2 dx$$

$$= \frac{\pi^2 n^2 T}{2\ell^2}D_n{}^2\cos^2(\omega_n t + \phi_n)\left[\int_0^\ell \cos^2\left[\frac{n\pi x}{\ell}\right]dx\right]$$

$$= \frac{\pi^2 n^2 T}{4\ell}D_n{}^2\cos^2(\omega_n t + \phi_n)$$

But

$$v = (T/\rho)^{1/2}$$

gives that

$$T = v^2\rho$$

and substituting this into the above expression for $U_n$ gives

$$E = K_n + U_n$$
$$= \frac{\pi^2 v^2 n^2 \rho}{4\ell} D_n^2 [\sin^2(\omega_n t + \phi_n) + \cos^2(\omega_n t + \phi_n)]$$
$$= \frac{\pi^2 v^2 n^2 \rho}{4\ell} D_n^2$$

If u(x,t) is given by

$$u(x,t) = \sum_{n=1}^{\infty} D_n \cos(\omega_n t + \phi_n) \sin(n\pi x/\ell)$$

then we have

$$\left[\frac{\partial u}{\partial y}\right] = \sum_{n=1}^{\infty} \frac{n\pi}{\ell} D_n \cos(\omega_n t + \phi_n) \cos(n\pi x/\ell)$$

and

$$\left[\frac{\partial u}{\partial t}\right] = -\sum_{n=1}^{\infty} \omega_n D_n \sin(\omega_n t + \phi_n) \sin(n\pi x/\ell)$$

$$U = \frac{1}{2} T \int_0^{\ell} \left[\frac{\partial u}{\partial x}\right]^2 dx$$

$$= \frac{1}{2} T \int_0^{\infty} \left[ \sum_{n=1}^{\infty} \frac{n\pi}{\ell} D_n \cos(\omega_n t + \phi_n) \cos(n\pi x/\ell) \times \sum_{m=1}^{\infty} \frac{m\pi}{\ell} D_m \cos(\omega_m t + \phi_m) \cos(m\pi x/\ell) \right] dx$$

$$= \frac{1}{2} T \sum_{n=1}^{\infty} \sum_{m=1}^{\infty} nm(\pi/\ell)^2 D_n D_m \cos(\omega_n t + \phi_n) \times \cos(\omega_m t + \phi_m) \times \int_0^{\ell} \cos(n\pi x/\ell) \cos(m\pi x/\ell)\, dx$$

Using the trigonometric identity

$$\cos\alpha\cos\beta = \frac{1}{2}[\cos(\alpha + \beta) + \cos(\alpha - \beta)]$$

we can show that the cross terms vanish.

Thus we have

$$U = \frac{T}{2} \sum_{n=1}^{\infty} \frac{n^2\pi^2}{\ell^2} D_n^2 \cos^2(\omega_n t + \phi_n)(\ell/2)$$

Substitute $T = v^2\rho$ into this to obtain

$$U = \frac{\pi^2 v^2 \rho}{4\ell}\sum_{n=1}^{\infty} D_n{}^2\cos^2(\omega_n t + \phi_n)$$

Similarly,

$$K = \frac{\rho}{2}\int_0^{\ell}\left[\frac{\partial u}{\partial t}\right]^2 dx$$

$$= \frac{\rho}{2}\int_0^{\ell}\left[\sum_{m=1}^{\infty} -\omega_n D_n \sin(\omega_m t + \phi_m)\sin(m\pi x/\ell)\right]^2 dx$$

Rewriting K as

$$K = \frac{\rho}{2}\sum_{n=1}^{\infty}\sum_{m=1}^{\infty}\Big[\omega_n\omega_m \sin(\omega_n t + \phi_n)\sin(\omega_m t + \phi_m) \times \int_0^{\ell}\sin(n\pi x/\ell)\sin(m\pi x/\ell)dx\Big]$$

and using the fact that

$$\int_0^{\ell}\sin(m\pi x/\ell)\sin(n\pi x/\ell)\ dx = \begin{cases} 0 \text{ if } m \neq n \\ \ell/2 \text{ if } m = n \end{cases}$$

we have

$$K = \frac{\rho\pi^2 v^2}{4\ell}\sum_{n=1}^{\infty} n^2 D_n{}^2\sin^2(\omega_n t + \phi_n)$$

Since $E = K + U$

$$E = \frac{\pi^2 v^2 \rho}{4\ell}\sum_{n=1}^{\infty} n^2 D_n{}^2[(\sin^2(\omega_n t + \phi_n) + \cos^2(\omega_n t + \phi_n)]$$

$$= \frac{\pi^2 v^2 \rho}{4\ell}\sum_{n=1}^{\infty} n^2 D_n{}^2$$

**3-31.** From Example 2-4, the solutions to the Schrödinger equation for a free particle are

$$\psi_1(x) = A_1 e^{ikx} = A_1\exp[i(2mE)^{1/2}x/\hbar]$$

and

$$\psi_2(x) = A_2 e^{-ikx} = A_2\exp[-i(2mE)^{1/2}x/\hbar]$$

If E is allowed to become negative, then in the case of $\psi_2(x)$

$$\psi_2(x) = A_2 \exp[(2mE)^{1/2}x/\hbar]$$

If x is allowed to become very large, then $\psi_2(x)$ becomes unbounded. Therefore, the energy must be positive. Using Eq. 3-11 for the momentum operator, we find that

$$\hat{P}_1\psi_1 = -i\hbar\frac{d\psi_1}{dx} = -i\hbar\frac{d}{dx}(A_1e^{ikx})$$
$$= -i\hbar(ikA_1e^{ikx})$$
$$= \hbar kA_1e^{ikx} = \hbar k\psi_1(x)$$

Likewise for $\psi_2$:

$$\hat{P}_2\psi_2 = -i\hbar\frac{d\psi_2}{dx} = -i\hbar\frac{d}{dx}(A_2e^{-ikx})$$
$$= -i\hbar(-ikA_2e^{-ikx})$$
$$= -\hbar kA_2e^{-ikx} = -\hbar k\psi_2(x)$$

Because all the energy of a free particle is kinetic energy, we use $E = p^2/2m$ and get

$$E = \frac{\hbar^2k^2}{2m}$$

We also have that

$$\psi_1^*(x)\psi_1(x) = (A_1^*e^{-ikx})(A_1e^{ikx})$$
$$= A_1^*A_1 = |A_1|^2 = \text{constant.}$$

Similarly,

$$\psi_2^*(x)\psi_2(x) = (A_2^*e^{ikx})(A_2e^{-ikx})$$
$$= A_2^*A_2 = |A_2|^2 = \text{constant.}$$

These last results say that the particle is equally likely to be found anywhere along the x axis. There is an infinite uncertainty in the location of the particle, but note that its momentum is sharp; there is no uncertainty in its momentum.

# CHAPTER FOUR

# THE POSTULATES AND GENERAL PRINCIPLES OF QUANTUM MECHANICS

**4-1.** We must determine which functions can be integrated when their magnitudes are squared (square integrable).

a)

$$\int_{-\infty}^{\infty} e^{-2x^2} dx = \left[\frac{\pi}{2}\right]^{1/2}$$

The normalized wave function is $(2/\pi)^{1/4} e^{-x^2}$

b)

$$\int_{0}^{\infty} e^{2x} dx \qquad \text{diverges. Not normalizable.}$$

c)

$$\int_{0}^{2\pi} (e^{i\theta})^* e^{i\theta} d\theta = \int_{0}^{2\pi} d\theta = 2\pi$$

The normalized wave function is $(2\pi)^{-1/2} e^{i\theta}$

d)

$$\int_{0}^{\infty} \sinh^2 x \, dx \qquad \text{diverges. Not normalizable.}$$

**4-2.** a) Not normalizable because

$$\int_{0}^{\infty} \frac{1}{x^2} dx \quad \text{is not finite.}$$

b) Acceptable.

c) Not normalizable because

$$\int_{-1}^{1} \frac{dx}{(1 - x^2)^2} \quad \text{is not finite.}$$

d) Acceptable

e) $\tan^{-1} x$ is not a single-valued function.

For example,

$$\tan^{-1}(1) = \pi/4, \ \pi/4 + 2\pi, \ \pi/4 + 4\pi, \text{ etc.}$$

**4-3.** First note that

$$\mathbf{i} \cdot \mathbf{i} = \mathbf{j} \cdot \mathbf{j} = \mathbf{k} \cdot \mathbf{k} = 1 \times 1 \times \cos 0 = 1$$

$$\mathbf{i} \cdot \mathbf{j} = 1 \times 1 \times \cos(\pi/2) = 0 = \mathbf{i} \cdot \mathbf{k} = \mathbf{j} \cdot \mathbf{k}$$

We now use these results to show that

$$\begin{aligned}\mathbf{a} \cdot \mathbf{b} &= (a_x\mathbf{i} + a_y\mathbf{j} + a_z\mathbf{k}) \cdot (b_x\mathbf{i} + b_y\mathbf{j} + b_z\mathbf{k}) \\ &= a_xb_x\mathbf{i}\cdot\mathbf{i} + a_xb_y\mathbf{i}\cdot\mathbf{j} + a_xb_z\mathbf{i}\cdot\mathbf{k} \\ &\quad + a_yb_x\mathbf{j}\cdot\mathbf{i} + a_yb_y\mathbf{j}\cdot\mathbf{j} + a_yb_z\mathbf{j}\cdot\mathbf{k} \\ &\quad + a_zb_x\mathbf{k}\cdot\mathbf{i} + a_zb_y\mathbf{k}\cdot\mathbf{j} + a_zb_z\mathbf{k}\cdot\mathbf{k} \\ &= a_xb_x + a_yb_y + a_zb_z\end{aligned}$$

Therefore, the dot product of $\mathbf{v}_1$ and $\mathbf{v}_2$ is

$$\mathbf{v}_1 \cdot \mathbf{v}_2 = 2 - 2 + 3 = 3$$

and

$$|\mathbf{v}_1| = (\mathbf{v}_1 \cdot \mathbf{v}_1)^{1/2} = \sqrt{4 + 1 + 9} = \sqrt{14}$$

$$|\mathbf{v}_2| = (\mathbf{v}_2 \cdot \mathbf{v}_2)^{1/2} = \sqrt{1 + 4 + 1} = \sqrt{6}$$

The angle between $\mathbf{v}_1$ and $\mathbf{v}_2$ is given by

$$\cos\theta = \frac{\mathbf{v}_1\cdot\mathbf{v}_2}{|\mathbf{v}_1||\mathbf{v}_2|} = \frac{3}{\sqrt{14}\ \sqrt{6}} = 0.327$$

$$\theta = \cos^{-1}0.327 = 70.89°$$

**4-4.** $\mathbf{i} \times \mathbf{i} = \mathbf{j} \times \mathbf{j} = \mathbf{k} \times \mathbf{k} = 0$ because $\theta = 0$ in the formula $\mathbf{a} \times \mathbf{b} = ab\sin\theta$.

$$\mathbf{i} \times \mathbf{j} = \mathbf{k} \text{ because } \theta = \pi/2,$$

$$|\mathbf{i}| = |\mathbf{j}| = 1$$

and the direction of $\mathbf{k}$ is perpendicular to the plane formed by $\mathbf{i}$ and $\mathbf{j}$.

$$\begin{aligned}\mathbf{a} \times \mathbf{b} &= (a_x\mathbf{i} + a_y\mathbf{j} + a_z\mathbf{k}) \times (b_x\mathbf{i} + b_y\mathbf{j} + b_z\mathbf{k}) \\ &= a_xb_x\mathbf{i}\times\mathbf{i} + a_yb_x\mathbf{j}\times\mathbf{i} + a_zb_x\mathbf{k}\times\mathbf{i} \\ &\quad + a_xb_y\mathbf{i}\times\mathbf{j} + a_yb_y\mathbf{j}\times\mathbf{j} + a_zb_y\mathbf{k}\times\mathbf{j} \\ &\quad + a_xb_z\mathbf{i}\times\mathbf{k} + a_yb_z\mathbf{j}\times\mathbf{k} + a_zb_z\mathbf{k}\times\mathbf{k} \\ &= 0 - a_yb_x\mathbf{i}\times\mathbf{j} - a_zb_x\mathbf{i}\times\mathbf{k} + a_xb_y\mathbf{i}\times\mathbf{j} \\ &\quad + 0 - a_zb_y\mathbf{j}\times\mathbf{k} + a_xb_z\mathbf{i}\times\mathbf{k} + a_yb_z\mathbf{j}\times\mathbf{k} + 0\end{aligned}$$

$$= (a_y b_z - a_z b_y)\mathbf{i} + (a_z b_x - a_x b_z)\mathbf{j}$$
$$+ (a_x b_y - a_y b_x)\mathbf{k}$$

This result is equivalent to

$$\mathbf{a} \times \mathbf{b} = \begin{vmatrix} \mathbf{i} & \mathbf{j} & \mathbf{k} \\ a_x & a_y & a_z \\ b_x & b_y & b_z \end{vmatrix}$$

To prove that $\mathbf{a} \times \mathbf{b} = -\mathbf{b} \times \mathbf{a}$, note that we are interchanging two rows of the above determinant. To prove that $\mathbf{a} \times \mathbf{a} = 0$, note that two rows are identical.

Lastly,

$$\mathbf{v}_1 \times \mathbf{v}_2 = \begin{vmatrix} \mathbf{i} & \mathbf{j} & \mathbf{k} \\ 2 & -1 & 3 \\ 1 & 2 & 1 \end{vmatrix} = -7\mathbf{i} + \mathbf{j} + 5\mathbf{k}$$

The angle between $\mathbf{v}_1$ and $\mathbf{v}_2$ is given by

$$\sin\theta = \frac{|\mathbf{v}_1 \times \mathbf{v}_2|}{|\mathbf{v}_1||\mathbf{v}_2|} = \frac{\sqrt{75}}{\sqrt{14}\ \sqrt{6}} = 0.945$$

$$\theta = 70.89°$$

**4-5.** Start with

$$\frac{df(t)}{dt} = \lim_{\Delta t \to 0} \frac{f(t + \Delta t) - f(t)}{\Delta t}$$

Then

$$\frac{d}{dt}\mathbf{a}\cdot\mathbf{b} = \lim_{\Delta t \to 0} \frac{\mathbf{a}(t + \Delta t)\cdot\mathbf{b}(t + \Delta t) - \mathbf{a}(t)\cdot\mathbf{b}(t)}{\Delta t}$$

$$= \lim_{\Delta t \to 0} \frac{[\mathbf{a}(t) + \Delta\mathbf{a}(t)]\cdot[\mathbf{b}(t) + \Delta\mathbf{b}(t)] - \mathbf{a}(t)\cdot\mathbf{b}(t)}{\Delta t}$$

$$= \lim_{\Delta t \to 0} \left[\frac{\Delta\mathbf{a}}{\Delta t}\cdot\mathbf{b}(t) + \mathbf{a}(t)\cdot\frac{\Delta\mathbf{b}}{\Delta t} + \text{higher-order terms}\right]$$

$$= \frac{d\mathbf{a}}{dt}\cdot\mathbf{b} + \mathbf{a}\cdot\frac{d\mathbf{b}}{dt}$$

Similarly

$$\frac{d}{dt}\mathbf{a} \times \mathbf{b} = \lim_{\Delta t \to 0} \frac{[\mathbf{a}(t) + \Delta\mathbf{a}(t)] \times [\mathbf{b}(t) + \Delta\mathbf{b}(t)] - \mathbf{a}(t) \times \mathbf{b}(t)}{\Delta t}$$

$$= \lim_{\Delta t \to 0}\left[\left[\frac{\Delta\mathbf{a}}{\Delta t} \times \mathbf{b}(t)\right] + \left[\mathbf{a}(t) \times \frac{\Delta\mathbf{b}}{\Delta t}\right] + \text{higher-order terms}\right]$$

$$= \frac{d\mathbf{a}}{dt} \times \mathbf{b} + \mathbf{a} \times \frac{d\mathbf{b}}{dt}$$

**4-6.** Start with

$$\mathbf{a} \times \mathbf{a} = 0$$

Then

$$\frac{d}{dt}(\mathbf{a} \times \mathbf{a}) = \left[\mathbf{a} \times \frac{d\mathbf{a}}{dt}\right] + \left[\frac{d\mathbf{a}}{dt} \times \mathbf{a}\right] = 0$$

**4-7.** Start with

$$m\frac{d^2\mathbf{r}}{dt^2} = \mathbf{F}$$

Operate with $\mathbf{r}\times$ to obtain

$$m\mathbf{r} \times \frac{d^2\mathbf{r}}{dt^2} = m\frac{d}{dt}\left[\mathbf{r} \times \frac{d\mathbf{r}}{dt}\right] = \mathbf{r} \times \mathbf{F}$$

or

$$\frac{d}{dt}(\mathbf{r} \times \mathbf{p}) = \underbrace{\mathbf{r} \times \mathbf{F}}_{\text{torque}}$$

If the torque equals zero, then $\frac{d}{dt}(\mathbf{r} \times \mathbf{p}) = 0$, and

$$\mathbf{r} \times \mathbf{p} = \text{angular momentum} = \text{constant.}$$

For circular motion, $\mathbf{r} \times \mathbf{p} = rmv\mathbf{u}$, where $\mathbf{u}$ is a unit vector perpendicular to the plane of the circular motion.

**4-8.**

$$\langle E \rangle = \int_0^a \Psi(x)\hat{H}\Psi(x)\,dx$$

$$= \int_0^a \Psi(x)\left[-\frac{\hbar^2}{2m}\frac{d^2}{dx^2}\right]\Psi(x)\,dx$$

$$= -\frac{630\hbar^2}{2ma^9}\int_0^a x^2(a - x)^2(2a^2 - 12ax + 12x^2)dx$$

To evaluate these integrals, it is most convenient to use

$$\int_0^1 x^m(1 - x)^n dx = \frac{m!n!}{(m + n + 1)!}$$

Thus

$$\langle E\rangle = -\frac{315\hbar^2}{ma^9}\left[2a^2\frac{2!2!a^5}{5!} - 12a\frac{3!2!a^6}{6!} + 12\frac{4!2!a^7}{7!}\right]$$

$$= -\frac{315\hbar^2}{ma^2}\left[-\frac{2}{7 \times 15}\right]$$

$$= \frac{6\hbar^2}{ma^2}$$

$$\langle E^2\rangle = \int_0^a \Psi(x)\hat{H}^2\Psi(x)dx$$

$$= \int_0^a \Psi(x)\left[\frac{\hbar^4}{4m^2}\frac{d^4}{dx^4}\right]\Psi(x)dx$$

$$= \frac{630\hbar^4}{4m^2a^9}\int_0^a x^2(a - x)^2(4\cdot3\cdot2)dx = \frac{126\hbar^4}{m^2a^4}$$

and

$$\sigma_E{}^2 = \langle E^2\rangle - \langle E\rangle^2 = \frac{90\hbar^4}{m^2a^4}$$

**4-9.** If the system is in one of its eigenstates, then

$$\hat{H}\psi_{n_xn_y} = E_{n_xn_y}\psi_{n_xn_y}$$

and

$$\hat{H}^2\psi_{n_xn_y} = E^2{}_{n_xn_y}\psi_{n_xn_y}$$

Therefore

$$\langle E\rangle = \iint\psi^*_{n_xn_y}\hat{H}\psi_{n_xn_y}dx\ dy = E_{n_xn_y}$$

$$\langle E^2\rangle = \iint\psi^*_{n_xn_y}\hat{H}^2\psi_{n_xn_y}dx\ dy = E^2{}_{n_xn_y}$$

and

$$\sigma_E{}^2 = \langle E^2\rangle - \langle E\rangle^2 = 0$$

**4-10.**

$$\langle \mathbf{p}\rangle = \int_0^a dx\int_0^b dy\,\psi_{n_xn_y}(x,y)\left[-i\hbar\left[\mathbf{i}\frac{\partial}{\partial x} + \mathbf{j}\frac{\partial}{\partial y}\right]\right]\psi_{n_xn_y}(x,y)$$

$$= -\frac{4i\hbar}{ab}\int_0^a dx\int_0^b dy\,\sin\frac{n_x\pi x}{a}\sin\frac{n_y\pi y}{b}$$

$$\times\left[\mathbf{i}\frac{n_x\pi}{a}\cos\frac{n_x\pi x}{a}\sin\frac{n_y\pi y}{b} + \mathbf{j}\frac{n_y\pi}{b}\sin\frac{n_x\pi x}{a}\cos\frac{n_y\pi y}{a}\right]$$

$$= 0$$

$$\langle p^2\rangle = \int_0^a dx\int_0^b dy\,\psi_{n_xn_y}(x,y)\left[-\hbar^2\left[\frac{\partial^2}{\partial x^2} + \frac{\partial^2}{\partial y^2}\right]\right] \times \psi_{n_xn_y}(x,y)$$

$$= \frac{4\hbar^2}{ab}\int_0^a dx\int_0^b dy\,\sin\frac{n_x\pi x}{a}\sin\frac{n_y\pi y}{b} \times\left[\frac{n_x{}^2\pi^2}{a^2} + \frac{n_y{}^2\pi^2}{b^2}\right]\sin\frac{n_x\pi x}{a}\sin\frac{n_y\pi y}{b}$$

$$= \pi^2\hbar^2\left[\frac{n_x{}^2}{a^2} + \frac{n_y{}^2}{b^2}\right] = \frac{h^2}{4}\left[\frac{n_x{}^2}{a^2} + \frac{n_y{}^2}{b^2}\right]$$

This result is the two-dimensional extension of Eq. 3-60,

$$\sigma_p{}^2 = \langle p^2\rangle - \langle p\rangle^2 = \langle p^2\rangle$$

**4-11.** To show that $\psi(x,y)$ is normalized

$$\int_0^a dx\int_0^b dy\,\psi^2(x,y) = \frac{900}{a^5b^5}\int_0^a dx\;x^2(a-x)^2\int_0^b dy\;y^2(b-y)^2$$

$$= \frac{900}{a^5b^5}\left[\frac{2\cdot 2a^5}{5!}\right]\left[\frac{2\cdot 2b^5}{5!}\right] = 1$$

We have used the integral

$$\int_0^1 x^n(1-x)^m dx = \frac{n!m!}{(n+m+1)!}$$

<E> is given by

$$\langle E\rangle = \frac{900}{a^5b^5}\int_0^a dx\int_0^b dy\ x(a-x)y(b-y) \times \left[-\frac{\hbar^2}{2m}\left[-2y(b-y) - 2x(a-x)\right]\right]$$

$$= \frac{900}{a^5b^5}\frac{\hbar^2}{m}\int_0^a dx\int_0^b dy\ x(a-x)y(b-y) \times [(y(b-y) + x(a-x)]$$

$$= \frac{900\hbar^2}{a^5b^5m}\left[\left[\frac{1\times 1a^3}{3!}\right]\left[\frac{2\times 2b^5}{5!}\right] + \left[\frac{2\times 2a^5}{5!}\right]\left[\frac{1\times 1b^3}{3!}\right]\right]$$

$$= \frac{5\hbar^2}{m}\left[\frac{1}{a^2} + \frac{1}{b^2}\right]$$

**4-12.**

$$S_1S_2 = \sum_{n=1}^{3}\ \sum_{m=1}^{2} a_nb_m = \left[\sum_{n=1}^{3} a_n\right]\left[\sum_{m=1}^{2} b_m\right]$$

$$= (a_1 + a_2 + a_3)(b_1 + b_2)$$

$$= a_1b_1 + a_1b_2 + a_2b_1 + a_2b_2 + a_3b_1 + a_3b_2$$

$$\neq a_1b_1 + a_2b_2$$

**4-13.** We must determine if the operator satisfies the condition

$$\int_{-\infty}^{\infty}\psi^*\hat{A}\psi dx = \int_{-\infty}^{\infty}(\hat{A}^*\psi^*)\psi dx$$

a)

$$\int_{-\infty}^{\infty}\psi^*\frac{d}{dx}\psi dx = \left|\psi^*\psi\right|_{-\infty}^{\infty} - \int_{-\infty}^{\infty}\psi\frac{d\psi^*}{dx}dx$$

$$= -\int_{-\infty}^{\infty}\psi\frac{d}{dx}\psi^* dx \qquad \text{Not Hermitian}$$

b) see page 125

c) see Example 4-4

d)

$$\int_{-\infty}^{\infty}\psi^* i\frac{d^2}{dx^2}\psi dx = \left|_{-\infty}^{\infty}\psi^* i\frac{d\psi}{dx}\right| - i\int_{-\infty}^{\infty}\frac{d\psi^*}{dx}\frac{d\psi}{dx}dx$$

$$= -i\left|_{-\infty}^{\infty}\psi\frac{d\psi^*}{dx}\right| + i\int_{-\infty}^{\infty}\psi\frac{d^2\psi^*}{dx^2}dx$$

$$= -\int_{-\infty}^{\infty}\psi\left[i\frac{d^2}{dx^2}\right]^*\psi^* dx \qquad \text{Not Hermitian}$$

e)

$$\int_{-\infty}^{\infty}\psi^* x\frac{d\psi}{dx}dx = \left|_{-\infty}^{\infty}\psi^* x\psi\right| - \int_{-\infty}^{\infty}\psi\left[x\frac{d\psi^*}{dx} + \psi^*\right]dx$$

$$\neq \int_{-\infty}^{\infty}\psi x\frac{d\psi^*}{dx}dx \qquad \text{Not Hermitian}$$

f)

$$\int_{-\infty}^{\infty}\psi^* x\psi dx = \int_{-\infty}^{\infty}\psi(x\psi)^* dx \qquad \text{Hermitian}$$

**4-14.** The operator is Hermitian, and so

$$\int g^*\hat{A}f dx = \int f\hat{A}^* g^* dx$$

Then

$$\int g^*(\hat{A} - \langle a\rangle)f dx = \int g^*\hat{A}f dx - \langle a\rangle\int g^* f dx$$

$$= \int f\hat{A}^* g^* dx - \langle a\rangle^*\int f g^* dx$$

$$= \int f(\hat{A} - \langle a\rangle)^* g^* dx$$

because $\langle a\rangle$ is real ($\hat{A}$ is Hermitian).

To prove that the sum of two Hermitian operators is real, consider

$$\int g^*(\hat{A} + \hat{B})f dx = \int g^*\hat{A}f dx + \int g^*\hat{B}f dx$$

$$= \int(\hat{A}^* g^*)f dx + \int(\hat{B}^* g^*)f dx$$

$$= \int f(\hat{A} + \hat{B})^* g^* dx$$

**4-15.** Start with

$$c_1^*c_2\int f^*\hat{A}g\,dx + c_2^*c_1\int g^*\hat{A}f\,dx$$

$$= c_1c_2^*\int f\hat{A}^*g^*dx + c_1^*c_2\int g\hat{A}^*f^*dx$$

$$c_1^*c_2\int(f^*\hat{A}g - g\hat{A}^*f^*)dx = c_1c_2^*\int(f\hat{A}^*g^* - g^*\hat{A}f)dx$$

If $z = x + iy$ and

$$z = x + iy = z^* = x - iy$$

then y must equal -y, which can be true only if y = 0, or if z is real. The integrals on the two sides of the above equation are complex conjugates of each other. Thus

$$\int f^*\hat{A}g\,dx = \int g\hat{A}^*f\,dx \qquad \text{(Eq. 4-29)}$$

**4-16.** If $\hat{A}$ is Hermitian, then

$$\int g^*\hat{A}f\,dx = \int(\hat{A}^*g^*)f\,dx$$

Let $f = \hat{B}\psi$ and $g = \psi$ and the problem is solved.

**4-17.** We have $\psi_n(x) = (2/a)^{1/2}\cos(n\pi x/a)$.

Then

$$\int_0^a \psi_n^*(x)\psi_m(x)dx = \frac{2}{a}\int_0^a \cos\frac{n\pi x}{a}\cos\frac{m\pi x}{a}dx$$

Use the trigonometric identity

$$\cos\alpha\cos\beta = \frac{1}{2}\cos(\alpha - \beta) + \frac{1}{2}\cos(\alpha + \beta)$$

to get

$$\frac{1}{a}\int_0^a \cos\frac{(n - m)\pi x}{a}dx + \frac{1}{a}\int_0^a \cos\frac{(n + m)\pi x}{a}dx$$

Unless n = m, these integrals are equal to zero because

$$\int_0^a \cos\frac{N\pi x}{a}dx = 0 \qquad N = 1,2,\ldots$$

If $m = n$, then

$$\frac{1}{a}\int_0^a dx = 1$$

and so we have proved that

$$\frac{2}{a}\int_0^a \cos\frac{n\pi x}{a}\cos\frac{n\pi x}{a}dx = \delta_{nm}$$

**4-18.** In the sum

$$\sum_{n=1}^{\infty} c_n\delta_{nm}$$

the $\delta_{nm}$ causes all the terms to be zero, except the term in which $n = m$. Thus

$$\sum_{n=1}^{\infty} c_n\delta_{nm} = c_m$$

Similarly, in the sum

$$\sum_n\sum_m a_nb_m\delta_{nm}$$

as we sum over $n$, all terms are zero except the term in which $n = m$. Thus,

$$\sum_n\sum_m a_nb_m\delta_{nm} = \sum_m a_mb_m = \sum_n a_nb_n$$

**4-19.** Let $f_0(x) = a_0 = 1$. Now require that

$$\int_0^1 f_0(x)f_1(x)dx = 0 = \int_0^1 (a_1 + b_1x)dx = a_1 + \frac{b_1}{2}$$

or

$$f_1(x) = a_1(1 - 2x)$$

To find $a_1$, require that

$$\int_0^1 f_1{}^2(x)dx = 1 = a_1{}^2\int_0^1 (1 - 2x)^2dx = \frac{1}{3}$$

or

$$f_1(x) = \sqrt{3}(1 - 2x)$$

Now require that

$$\int_0^1 f_0(x)f_2(x)dx = 0 = a_2 + \frac{b_2}{2} + \frac{c_2}{3}$$

and

$$\int_0^1 f_1(x)f_2(x)dx = \sqrt{3}\int_0^1 (1 - 2x)(a_2 + b_2x + c_2x^2)dx$$

$$= 0$$

$$= \sqrt{3}\left[a_2 + \frac{b_2}{2} + \frac{c_2}{3} - a_2 - \frac{2}{3}b_2 - \frac{c_2}{2}\right]$$

or

$$a_2 + \frac{2}{3}b_2 + \frac{c_2}{2} = 0$$

Solving the two equations for $a_2$, $b_2$ and $c_2$, we obtain

$$b_2 = -c_2 = -6a_2$$

Thus

$$f_2(x) = a_2(1 - 6x + 6x^2)$$

The value of $a_2$ can be found by requiring that

$$\int_0^1 f_2^2(x)dx = 1$$

giving

$$f_2(x) = \sqrt{5}(1 - 6x + 6x^2)$$

**4-20.** Start with

$$f(x) = \sum_{n=1}^{\infty} b_n \sin\frac{n\pi x}{a}$$

Multiply both sides by $\sin(m\pi x/a)$ and integrate from 0 to a

$$\int_0^a f(x)\sin\frac{m\pi x}{a}dx = \sum_{n=1}^{\infty} b_n \int_0^a \sin\frac{n\pi x}{a}\sin\frac{m\pi x}{a}dx$$

$$= \sum_{n=1}^{\infty} b_n \frac{a}{2}\delta_{nm} = \frac{a}{2}b_m$$

or

$$b_m = \frac{2}{a}\int_0^a f(x)\sin\frac{m\pi x}{a}dx$$

If $f(x) = x$, then

$$b_m = \frac{2}{a}\int_0^a x \sin\frac{m\pi x}{a}dx = \frac{2}{a}\left[- \frac{a^2}{m\pi}\cos m\pi\right]$$

$$= - \frac{2a}{m\pi}(-1)^m = \frac{2a}{m\pi}(-1)^{m+1}$$

Thus,

$$x = \frac{2a}{\pi}\sum_{m=1}^{\infty}\frac{(-1)^{m+1}}{m}\sin\frac{m\pi x}{a}$$

**4-21.** A plot of $\sin(n\pi x/a)$ is given in Figure 3-2. Note that $\sin(n\pi x/a)$ is even about $a/2$ when $n$ is even and is odd about $a/2$ when $n$ is odd.

The integral leading to Eq. 4-49 is

$$\int_0^a x(a - x)\sin\frac{n\pi x}{a}dx$$

Because $x(a - x)$ is even about $a/2$, the integral will vanish if $\sin(n\pi x/a)$ is odd.

**4-22.** We are given

$$T(x,t) = \sum_{n=1}^{\infty} B_n \exp(-\lambda_n^2 t)\sin\frac{n\pi x}{\ell}$$

Set $t = 0$, multiply by $\sin(m\pi x/\ell)$ and integrate from 0 to $\ell$ (see Problem 4-20)

$$B_n = \frac{2}{\ell}\int_0^{\ell} T(x,0)\sin\frac{n\pi x}{\ell}dx$$

$$= \frac{2}{\ell}\left[\int_0^{\ell/2} x \sin\frac{n\pi x}{\ell}dx + \int_{\ell/2}^{\ell} (\ell - x)\sin\frac{n\pi x}{\ell}dx\right]$$

$$= \frac{2}{\ell}\left[ \left|_{0}^{\ell/2} \frac{\ell^2\sin(n\pi x/\ell)}{n^2\pi^2} - \frac{\ell x\cos(n\pi x/\ell)}{n\pi} \right| \right.$$

$$+ \ \ell \left|_{\ell/2}^{\ell} - \frac{\ell\cos(n\pi x/\ell)}{n\pi} \right|$$

$$\left. - \left|_{\ell/2}^{\ell} \frac{\ell^2\sin(n\pi x/\ell)}{n^2\pi^2} - \frac{\ell x\cos(n\pi x/\ell)}{n\pi} \right| \right]$$

$$= \frac{2}{\ell}\left[\frac{2\ell^2}{n^2\pi^2}\sin\frac{n\pi}{2}\right] = \frac{4\ell}{\pi^2}\left[\frac{\sin(n\pi/2)}{n^2}\right]$$

Therefore,

$$T(x,t) = \frac{4\ell}{\pi^2}\left[e^{-\lambda_1^2 t}\sin\frac{\pi x}{\ell} - \frac{1}{9}e^{-\lambda_3^2 t}\sin\frac{3\pi x}{\ell} + \cdots\right]$$

**4-23.** Multiply

$$\Psi(x,t) = \sum_n C_n(t)\phi_n(x)$$

by $\phi_m^*(x)$ and integrate to obtain

$$\int_{-\infty}^{\infty}\phi_m^*(x)\Psi(x,t)dx = \sum_n C_n(t)\int_{-\infty}^{\infty}\phi_m^*(x)\phi_n(x)dx$$

$$= \sum_n C_n(t)\delta_{nm} = C_m(t)$$

Substitute Eq. 4-21

$$\Psi(x,t) = \sum_n c_n\psi_n(x)e^{-iE_nt/\hbar}$$

into the above integral to obtain

$$C_m(t) = \sum_n c_n\exp(-iE_nt/\hbar)\int_{-\infty}^{\infty}\phi_m^*(x)\psi_n(x)dx$$

If $\Psi(x,0) = \phi_k(x)$, then

$$C_m(0) = \int_{-\infty}^{\infty}\phi_m^*(x)\Psi(x,0)dx$$

$$= \int_{-\infty}^{\infty}\phi_m^*(x)\phi_k(x)dx$$

$$= \delta_{mk}$$

and

$$P_m(0) = |C_m(0)|^2 = \delta_{mk}$$

If $\hat{A}$ and $\hat{H}$ have different eigenvalues, then $P_m(t) = |C_m(t)|^2$ varies with time. The quantity $P_k(t)$ decreases with time from its initial value of unity, and $P_m(t)$ $(m \neq k)$ increases with time from an initial value of zero. If $\hat{A}$ and $\hat{H}$ have a mutual set of eigenfunctions, then $C_k(t) = c_k \exp(-iE_k t/\hbar)$ and so $P_k(t) = |C_k(t)|^2 = |c_k|^2$ is independent of time.

**4-24.** We must evaluate

$$[\hat{A},\hat{B}] = \hat{A}\hat{B} - \hat{B}\hat{A}$$

a) $$\hat{A}\hat{B}f = \frac{d^2}{dx^2}xf = \frac{d}{dx}\left[f + x\frac{df}{dx}\right]$$

$$= 2\frac{df}{dx} + x\frac{d^2f}{dx^2} = \left[2\frac{d}{dx} + x\frac{d^2}{dx^2}\right]f$$

$$\hat{B}\hat{A}f = x\frac{d^2}{dx^2}f$$

$$\hat{A}\hat{B} - \hat{B}\hat{A} = 2\frac{d}{dx}$$

b) $$\hat{A}\hat{B}f = \left[\frac{d}{dx} - x\right]\left[\frac{df}{dx} + xf\right]$$

$$= \frac{d^2f}{dx^2} + f - x^2f = \left[\frac{d^2}{dx^2} + 1 - x^2\right]f$$

$$\hat{B}\hat{A}f = \left[\frac{d}{dx} + x\right]\left[\frac{df}{dx} - xf\right]$$

$$= \frac{d^2f}{dx^2} - f - x^2f = \left[\frac{d^2}{dx^2} - 1 - x^2\right]f$$

$$\hat{A}\hat{B} - \hat{B}\hat{A} = 2$$

c)

$$\hat{A}\hat{B}f = \int_0^x dx'\frac{df}{dx'} = f(x) - f(0)$$

$$\hat{B}\hat{A}f = \frac{d}{dx}\int_0^x dx' f(x') = f(x)$$

$$[\hat{A}\hat{B} - \hat{B}\hat{A}]f(x) = -f(0)$$

d) $$\hat{A}\hat{B}f = \left[\frac{d^2}{dx^2} - x\right]\left[\frac{df}{dx} + x^2 f\right]$$

$$= \frac{d^3f}{dx^3} + x^2\frac{d^2f}{dx^2} + 3x\frac{df}{dx} + (2 - x^3)f$$

$$\hat{B}\hat{A}f = \left[\frac{d}{dx} + x^2\right]\left[\frac{d^2f}{dx^2} - xf\right]$$

$$= \frac{d^3f}{dx^3} + x^2\frac{d^2f}{dx^2} - x\frac{df}{dx} - (1 + x^3)f$$

$$[\hat{A},\hat{B}] = \hat{A}\hat{B} - \hat{B}\hat{A} = 4x\frac{d}{dx} + 3$$

**4-25.** $$\hat{L}_x\hat{L}_y f = \left[-i\hbar\left[y\frac{\partial}{\partial z} - z\frac{\partial}{\partial y}\right]\right]\left[-i\hbar\left[z\frac{\partial}{\partial x} - x\frac{\partial}{\partial z}\right]\right]f$$

$$= -\hbar^2\left[y\frac{\partial}{\partial z} - z\frac{\partial}{\partial y}\right]\left[z\frac{\partial f}{\partial x} - x\frac{\partial f}{\partial z}\right]$$

$$= -\hbar^2\left[y\frac{\partial f}{\partial x} + yz\frac{\partial^2 f}{\partial z\,\partial x} - xy\frac{\partial^2 f}{\partial z^2} - z^2\frac{\partial^2 f}{\partial x\,\partial y} + xz\frac{\partial^2 f}{\partial y\,\partial z}\right]$$

$$\hat{L}_y\hat{L}_x f = \left[-i\hbar\left[z\frac{\partial}{\partial x} - x\frac{\partial}{\partial z}\right]\right]\left[-i\hbar\left[y\frac{\partial}{\partial z} - z\frac{\partial}{\partial y}\right]\right]f$$

$$= -\hbar^2\left[z\frac{\partial}{\partial x} - x\frac{\partial}{\partial z}\right]\left[y\frac{\partial f}{\partial z} - z\frac{\partial f}{\partial y}\right]$$

$$= -\hbar^2\left[yz\frac{\partial^2 f}{\partial x\,\partial z} - z^2\frac{\partial^2 f}{\partial x\,\partial y} - xy\frac{\partial^2 f}{\partial z^2} + x\frac{\partial f}{\partial y} + xz\frac{\partial^2 f}{\partial z\,\partial y}\right]$$

$$[\hat{L}_x,\hat{L}_y] = -\hbar^2\left[y\frac{\partial}{\partial x} - x\frac{\partial}{\partial y}\right] = i\hbar\left[-i\hbar\left[y\frac{\partial}{\partial x} - x\frac{\partial}{\partial y}\right]\right]$$

$$= i\hbar\hat{L}_z$$

The other relations are proved in a similar manner. The pattern involves the cyclic permutation of x,y and z.

To prove that $\hat{L}^2$ commutes with $\hat{L}_x$, $\hat{L}_y$ and $\hat{L}_z$, let's use $\hat{L}_x$ as an example. Clearly $\hat{L}_x{}^2$ commutes with $\hat{L}_x$. The other terms are given by

$$\hat{L}_y{}^2\hat{L}_x - \hat{L}_x\hat{L}_y{}^2 = \hat{L}_y{}^2\hat{L}_x - (\hat{L}_y\hat{L}_x + i\hbar\hat{L}_z)\hat{L}_y$$

and

$$\hat{L}_z{}^2\hat{L}_x - \hat{L}_x\hat{L}_z{}^2 = \hat{L}_z{}^2\hat{L}_x - (\hat{L}_z\hat{L}_x - i\hbar\hat{L}_y)\hat{L}_z$$

We now add these two lines to obtain

$$\begin{aligned}
&\hat{L}_y{}^2\hat{L}_x + \hat{L}_z{}^2\hat{L}_x - \hat{L}_y\hat{L}_x\hat{L}_y - \hat{L}_z\hat{L}_x\hat{L}_z + i\hbar(i\hbar\hat{L}_x) \\
&\quad = \hat{L}_y(\hat{L}_y\hat{L}_x - \hat{L}_x\hat{L}_y) + \hat{L}_z(\hat{L}_z\hat{L}_x - \hat{L}_x\hat{L}_z) - \hbar^2\hat{L}_x \\
&\quad = \hat{L}_y(-i\hbar\hat{L}_z) + \hat{L}_z(i\hbar\hat{L}_y) - \hbar^2\hat{L}_x \\
&\quad = i\hbar(\hat{L}_z\hat{L}_y - \hat{L}_y\hat{L}_z) - \hbar^2\hat{L}_x \\
&\quad = i\hbar(-i\hbar\hat{L}_x) - \hbar^2\hat{L}_x = 0
\end{aligned}$$

**4-26.** Start with $\int\phi_1{}^*\hat{A}\hat{B}\phi_2 dx$. Let $g^* = \phi_1{}^*$ and $f = \hat{B}\phi_2$ in Eq. 4-29 to obtain

$$\int\phi_1{}^*\hat{A}\hat{B}\phi_2 dx = \int(\hat{A}\phi_1)^*\hat{B}\phi_2 dx$$

Now let $(\hat{A}\phi_1)^* = g^*$ and use Eq. 4-29 again

$$\int(\hat{A}\phi_1)^*\hat{B}\phi_2 dx = \int\phi_2(\hat{B}\hat{A}\phi_1)^* dx$$

Thus we have

$$\int\phi_1{}^*\hat{A}\hat{B}\phi_2 dx = \int\phi_2(\hat{B}\hat{A}\phi_1)^* dx$$

Notice that $\hat{A}\hat{B}$ is Hermitian only if $\hat{A}$ and $\hat{B}$ commute.

**4-27.** Consider the Taylor series of $e^{\hat{A}+\hat{B}}$

$$e^{\hat{A}+\hat{B}} = \hat{I} + (\hat{A} + \hat{B}) + \frac{1}{2!}(\hat{A} + \hat{B})^2 + \frac{1}{3!}(\hat{A} + \hat{B})^3 + \cdots)$$

Let's look at the quadratic term.

$$(\hat{A} + \hat{B})^2 = (\hat{A} + \hat{B})(\hat{A} + \hat{B}) = \hat{A}^2 + \hat{A}\hat{B} + \hat{B}\hat{A} + \hat{B}^2$$

Now consider

$$\begin{aligned}
e^{\hat{A}}e^{\hat{B}} &= (\hat{I} + \hat{A} + \tfrac{1}{2}\hat{A}^2 + \cdots)(I + \hat{B} + \tfrac{1}{2}\hat{B}^2 + \cdots) \\
&= I + \hat{A} + \hat{B} + \hat{A}\hat{B} + \tfrac{1}{2}\hat{A}^2 + \tfrac{1}{2}\hat{B}^2 + \cdots \\
&= I + \hat{A} + \hat{B} + \tfrac{1}{2}(\hat{A}^2 + 2\hat{A}\hat{B} + \hat{B}^2) + \cdots
\end{aligned}$$

The two expansions will be the same only if $\hat{A}$ and $\hat{B}$ commute.

**4-28.** If $[\hat{A},\hat{H}] = 0$, then the $\{\phi_n(x)\}$ and the $\{\psi_m(x)\}$ are the same as in Problem 4-23. In this case,

$$\int_{-\infty}^{\infty} \phi_m^*(x)\psi_n(x)dx = \delta_{mn}$$

and

$$C_m(t) = C_m \exp(-iE_n t/\hbar)$$

and

$$P_m(t) = |C_m(t)|^2 = c_m^* c_m$$

is independent of time. Thus a is a conserved quantity.

**4-29.** First consider

$$\left[\int (f^*g + fg^*)dx\right]^2 = \left[\int f^*gdx + \int fg^*dx\right]^2$$

Let $\int f^*gdx = a + ib$. Then $\int f^*gdx + \int fg^*dx = 2a$, and so

$$\left[\int (f^*g + fg^*)dx\right]^2 = 4\left|\int f^*gdx\right|^2 = 4\left|\int fg^*dx\right|^2$$

Using the first inequality given in the problem, we have

$$\left[\int |f|^2dx\right]\left[\int |g|^2dx\right] \geq \left|\int f^*gdx\right|^2$$

Now consider the intrinsically positive quantity

$$\int (f+\lambda g)^*(f+\lambda g)dx$$
$$= |\lambda|^2\int g^*gdx + \lambda^*\int fg^*dx + \lambda\int f^*gdx + \int f^*fdx$$
$$\geq 0$$

and let

$$\lambda = -\frac{\int g^*fdx}{\int g^*gdx} = -\frac{\left[\int gf^*dx\right]^*}{\int g^*gdx}$$

Then

$$\frac{\left|\int g^*f dx\right|^2}{\int g^*g dx} - \frac{\int f^*g dx \int fg^* dx}{\int g^*g dx} - \frac{\int f^*g dx \int g^*f dx}{\int g^*g dx}$$

$$+ \int f^*f dx \geq 0$$

or upon cancellation

$$\int f^*f dx \geq \frac{\left|\int f^*g dx\right|^2}{\int g^*g dx}$$

**4-30.** Differentiate $\langle x\rangle = \int \Psi^*(x,t)\hat{X}\Psi(x,t)dx$ to obtain

$$\frac{d\langle x\rangle}{dt} = \int \frac{\partial\Psi^*}{\partial t}\hat{X}\Psi dx + \int \Psi^*\hat{X}\frac{\partial\Psi}{\partial t}dx$$

$$= -\frac{1}{i\hbar}\int(\hat{H}\Psi)^*\hat{X}\Psi dx + \frac{1}{i\hbar}\int\Psi^*\hat{X}(\hat{H}\Psi)dx$$

$$= -\frac{1}{i\hbar}\int\Psi^*\hat{H}\hat{X}\Psi dx + \frac{1}{i\hbar}\int\Psi^*\hat{X}(\hat{H}\Psi)dx$$

$$= \frac{1}{i\hbar}\int\Psi^*(\hat{X}\hat{H} - \hat{H}\hat{X})\Psi dx$$

$$= \frac{i}{\hbar}\int\Psi^*(\hat{H}\hat{X} - \hat{X}\hat{H})\Psi dx$$

$$(\hat{H}\hat{X} - \hat{X}\hat{H})f = \left[-\frac{\hbar^2}{2m}\frac{d^2}{dx^2} + U(x)\right]xf$$

$$- x\left[-\frac{\hbar^2}{2m}\frac{d^2f}{dx^2} + U(x)f\right]$$

$$= -2\frac{\hbar^2}{2m}\frac{df}{dx}$$

$$= -\frac{\hbar^2}{m}\frac{i}{\hbar}\hat{P}_x = -\frac{i\hbar}{m}\hat{P}_x$$

Finally

$$\frac{d\langle x\rangle}{dt} = \frac{i}{\hbar}\int\Psi^*[\hat{H},\hat{X}]\Psi dx$$

$$= \frac{1}{m}\int\Psi^*\hat{P}_x\Psi dx = \frac{\langle P_x\rangle}{m}$$

**4-31.** If

$$\langle F\rangle = \int\Psi^*(x,t)\hat{F}\Psi(x,t)\,dx$$

then

$$\frac{d\langle F\rangle}{dt} = \int\frac{\partial\Psi^*}{\partial t}\hat{F}\Psi(x,t)\,dx + \int\Psi^*(x,t)\hat{F}\frac{\partial\Psi}{\partial t}dx$$

$$= -\frac{1}{i\hbar}\int(\hat{H}^*\Psi^*)\hat{F}\Psi dx + \frac{1}{i\hbar}\int\Psi^*\hat{F}\hat{H}\Psi dx$$

$$= -\frac{1}{i\hbar}\int\Psi^*\hat{H}\hat{F}\Psi dx + \frac{1}{i\hbar}\int\Psi^*\hat{F}\hat{H}\Psi dx$$

$$= \int\Psi^*\frac{i}{\hbar}(\hat{H}\hat{F} - \hat{F}\hat{H})\Psi dx$$

Let $\hat{F} = \hat{P}_x$,

$$\hat{H}\hat{P}_x - \hat{P}_x\hat{H} = U(x)\hat{P}_x - \hat{P}_xU(x)$$

$$= i\hbar\frac{dU}{dx}$$

Thus

$$\frac{d\langle P_x\rangle}{dt} = -\int\Psi^*\frac{dU}{dx}\Psi dx = \left\langle -\frac{dU}{dx}\right\rangle$$

# CHAPTER FIVE

# THE HARMONIC OSCILLATOR

**5-1.** Newton's equation for a harmonic oscillator is

$$\frac{d^2x}{dt^2} + \frac{k}{m}x = 0$$

Using $x(t) = A\sin\omega t + B\cos\omega t$, we find that

$$-A\omega^2\sin\omega t - B\omega^2\cos\omega t + \frac{k}{m}(A\sin\omega t + B\cos\omega t) = 0$$

if we recall that $\omega^2 = k/m$.

**5-2.** Newton's equation for a harmonic oscillator is

$$\frac{d^2x}{dt^2} + \frac{k}{m}x = 0$$

Using $x(t) = C\sin(\omega t + \phi)$, we find that

$$-C\omega^2\sin(\omega t + \phi) + \frac{k}{m}C\sin(\omega t + \phi) = 0$$

if we recall that $\omega^2 = k/m$.

**5-3.** Consider the trajectory

$$x(t) = C\sin(\omega t + \phi)$$

Because the sine function varies from +1 to −1, the value of $x$ varies from $+C$ to $-C$. To find the period of oscillation, we wish to determine the smallest nonzero value of $\tau$ that satisfies the condition

$$\begin{aligned}\sin(\omega t + \phi) &= \sin[\omega(t + \tau) + \phi] \\ &= \sin[\omega t + \phi + \omega\tau] \\ &= \sin(\omega t + \phi)\cos\omega\tau \\ &\qquad + \cos(\omega t + \phi)\sin\omega\tau \\ &= \sin(\omega t + \phi)\end{aligned}$$

Thus, the trajectory repeats itself when $\tau$ satisfies the two conditions

$$\cos\omega\tau = 1 \qquad \sin\omega\tau = 0$$

or when

$$\omega\tau = 2\pi n \qquad n = 1, 2, \ldots$$

The smallest value of $\tau$ is

$$\tau = \frac{2\pi}{\omega}$$

which is the period of the oscillation. This is the time it takes for the oscillator to undergo one cycle. The frequency of oscillation is

$$\nu = \frac{1}{\tau} = \frac{\omega}{2\pi}$$

**5-4.** The value of $\langle K\rangle$ is

$$\langle K\rangle = \frac{1}{\tau}\int_0^{\tau} \frac{m\omega^2C^2}{2}\cos^2(\omega t + \phi)dt$$

$$= \frac{m\omega C^2}{2\tau}\int_{\phi}^{\omega\tau+\phi} \cos^2 x\,dx$$

$$= \frac{m\omega C^2}{2\tau}\left|\frac{x}{2} + \sin 2x\right|_{\phi}^{\omega\tau+\phi}$$

$$= \frac{m\omega C^2}{2\tau}\left[\frac{\omega\tau}{2} + \frac{\sin 2(\omega\tau + \phi) - \sin 2\phi}{4}\right]$$

$$= \frac{m\omega C^2}{2\tau}\left[\frac{\omega\tau}{2} + \frac{(\sin 2\omega\tau)(\cos 2\phi) + (\cos 2\omega\tau)(\sin 2\phi) - \sin 2\phi}{4}\right]$$

$$= \frac{m\omega^2C^2}{4} = \frac{kC^2}{4} = \frac{E}{2}$$

where we have used the fact that $\tau = 2\pi/\omega$ in evaluating the trigonometric terms.

Similarly

$$\langle U\rangle = \frac{kC^2}{2\tau}\int_0^{\tau}\sin^2(\omega t + \phi)dt$$

$$= \frac{kC^2}{2\tau\omega}\left[\frac{\omega\tau}{2}\right] = \frac{kC^2}{4} = \frac{E}{2}$$

The natural motion of a harmonic oscillator is such that $\langle K\rangle = \langle U\rangle$ over a cycle of motion.

5-5.

a)
$$m\frac{d^2x}{dt^2} = mg$$

Integration gives

$$x(t) = \frac{1}{2}gt^2 + v_ot$$

The instantaneous kinetic energy and potential energy is.

$$K(t) = \frac{m}{2}\dot{x}^2 = \frac{mv_o^2}{2} + mgv_ot + \frac{mg^2t^2}{2}$$

$$U(t) = mg(x_o - x) = mgx_o - mgv_ot - \frac{mg^2t^2}{2}$$

$$K(t) + U(t) = \frac{mv_o^2}{2} + mgx_o = \text{initial total energy}$$

b)
$$m\frac{d^2x}{dt^2} = -mg$$

Integration gives

$$x(t) = -\frac{1}{2}gt^2 - v_ot + x_o$$

The instantaneous kinetic energy and potential energy is

$$K(t) = \frac{m}{2}\dot{x}^2 = \frac{mg^2t^2}{2} + mgv_ot + \frac{m}{2}v_o^2$$

$$U(t) = mgx = mgx_o - \frac{mg^2t^2}{2} - mgv_ot$$

$$K(t) + U(t) = \frac{mv_o^2}{2} + mgx_o = \text{initial total energy}$$

5-6. The forces acting upon particles 1 and 2 are

$$f_1 = -\frac{\partial U}{\partial x_1} \qquad f_2 = -\frac{\partial U}{\partial x_2} = \frac{\partial U}{\partial x_1} = -f_1$$

The center-of-mass and relative coordinates are

$$X = \frac{m_1x_1 + m_2x_2}{M} \qquad x = x_1 - x_2$$

Multiply the first of these equations by M and the second by $m_2$ and add to obtain

$$x_1 = X + \frac{m_2}{M}x$$

Multiply the first by M and the second by $m_1$ and subtract to obtain

$$x_2 = X - \frac{m_1}{M}x$$

Substitute these expressions for $x_1$ and $x_2$ into Newton's equations

$$m_1\frac{d^2x_1}{dt^2} = m_1\frac{d^2X}{dt^2} + \frac{m_1m_2}{M}\frac{d^2x}{dt^2} = -\frac{\partial U}{\partial x_1} = -\frac{\partial U}{\partial x}$$

$$m_2\frac{d^2x_2}{dt^2} = m_2\frac{d^2X}{dt^2} - \frac{m_1m_2}{M}\frac{d^2x}{dt^2} = -\frac{\partial U}{\partial x_2} = \frac{\partial U}{\partial x}$$

We add these two equations to obtain

$$(m_1 + m_2)\frac{d^2X}{dt^2} = 0$$

which says that the center of mass moves uniformly. We now divide the first of Newton's equations by $m_1$ and the second by $m_2$ and subtract to obtain

$$\frac{d^2x}{dt^2} = - \left[\frac{1}{m_1} + \frac{1}{m_2}\right]\frac{\partial U}{\partial x}$$

or

$$\mu\frac{d^2x}{dt^2} = - \frac{\partial U}{\partial x}$$

This is the equation of motion of one body of mass $\mu$ moving under a force $-\partial U/\partial x$.

**5-7.** The equations in the x-direction are the same as in problem 5-6. Let's look at the y-direction.

$$m_1\frac{d^2y_1}{dt^2} = - \frac{\partial U}{\partial y_1} \qquad\qquad m_2\frac{d^2y_2}{dt^2} = - \frac{\partial U}{\partial y_2}$$

In terms of the y-component of the center-of-mass and relative coordinates

$$y_1 = Y + \frac{m_2}{M}y \qquad\qquad y_2 = Y - \frac{m_1}{M}y$$

Newton's equations become

$$m_1\frac{d^2Y}{dt^2} + \frac{m_1m_2}{M}\frac{d^2y}{dt^2} = - \frac{\partial U}{\partial y}$$

$$m_2\frac{d^2Y}{dt^2} - \frac{m_1m_2}{M}\frac{d^2y}{dt^2} = \frac{\partial U}{\partial y}$$

Following the manipulations in Problem 5-6, we can get

$$M\frac{d^2Y}{dt^2} = 0$$

and

$$\mu\frac{d^2y}{dt^2} = -\frac{\partial U}{\partial y}$$

It is straightforward to derive similar equations for the z coordinates, and so we have

$$M\frac{d^2\mathbf{R}}{dt^2} = 0$$

and

$$\mu\frac{d^2\mathbf{r}}{dt^2} = -\nabla U$$

as the three-dimensional generalization of Problem 5-6.

**5-8.** Expand U(x) in a Taylor series about x = 0

$$U(x) = U(0) + \left[\frac{dU}{dx}\right]_{x=0} x + \frac{1}{2}\left[\frac{d^2U}{dx^2}\right]_{x=0} x^2 + \frac{1}{3!}\left[\frac{d^3U}{dx^3}\right]_{x=0} x^3 + \cdots$$

It is straightforward to show that

$$\left[\frac{dU}{dx}\right]_{x=0} = 0$$

that

$$\left[\frac{d^2U}{dx^2}\right]_{x=0} = 2D_e\beta^2$$

and that

$$\left[\frac{d^3U}{dx^3}\right]_{x=0} = -6D_e\beta^3$$

Therefore

$$U(x) = D_e\beta^2x^2 - D_e\beta^3x^3 + \cdots$$

Comparing this result to $U(x) = (1/2)kx^2$, we find that

$$k = 2D_e\beta^2 = 2(7.31 \times 10^{-19}\text{J})(1.82 \times 10^{10}\ \text{m}^{-1})^2$$
$$= 484\ \text{N}\cdot\text{m}^{-1}$$

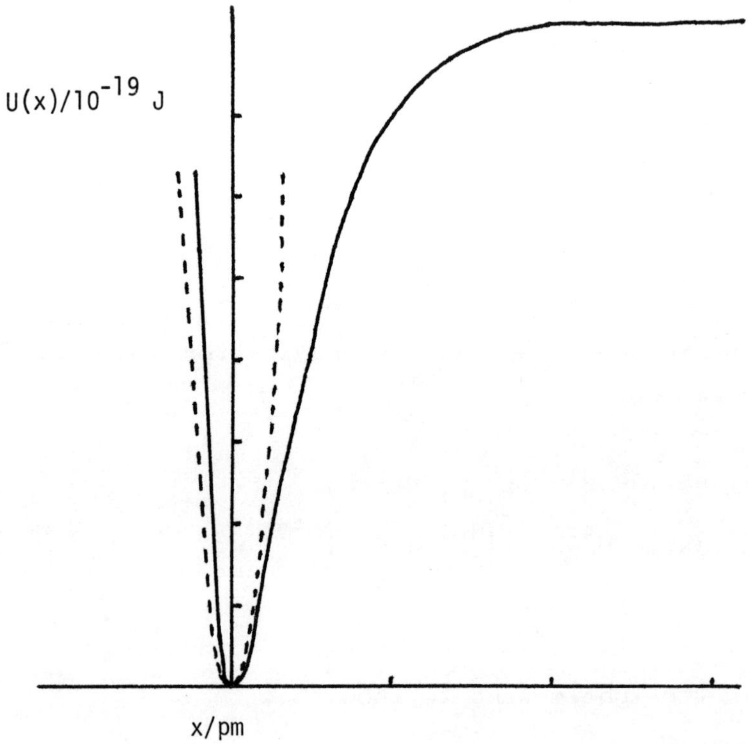


**5-9.** Use Eq. 5-40

$$\bar{\nu}_{obs} = \frac{1}{2\pi c}\left[\frac{k}{\mu}\right]^{1/2}$$

or

$$k = (2\pi c\bar{\nu}_{obs})^2\mu$$

$$= [2\pi(2.998 \times 10^{10}\ \text{cm}\cdot\text{s}^{-1})(2.60 \times 10^{3}\ \text{cm}^{-1})]^2$$

$$\times [(1 \times 79)/80\ \text{amu}]$$

$$\times [1.6607 \times 10^{-27}\ \text{kg}\cdot\text{amu}^{-1}]$$

$$= 393\ \text{N}\cdot\text{m}^{-1}$$

The period of vibration is given by

$$\tau = \frac{1}{\nu_{obs}} = \frac{1}{c\bar{\nu}_{obs}} = 1.28 \times 10^{-14}\text{s}$$

**5-10.** Use Eq. 5-40

$$\bar{\nu}_{obs} = \frac{1}{2\pi c}\left[\frac{k}{\mu}\right]^{1/2}$$

$$= \frac{1}{2\pi(2.998 \times 10^{10}\ \text{cm}\cdot\text{s}^{-1})}$$

$$\times \left[\frac{240\text{N}\cdot\text{m}^{-1}}{((79/2)\text{amu})(1.66 \times 10^{-27}\ \text{kg}\cdot\text{amu}^{-1})}\right]^{1/2}$$

$$= 321\ \text{cm}^{-1}$$

The zero-point energy is given by

$$E_0 = \frac{1}{2}h\nu = \frac{1}{2}hc\bar{\nu}_{obs} = 3.19 \times 10^{-21}\ \text{J}$$

**5-11.** The Schrödinger equation for a harmonic oscillator is

$$\frac{d^2\psi_n}{dx^2} + \frac{2\mu}{\hbar^2}(E_n - \frac{1}{2}kx^2)\psi_n(x) = 0$$

From Eqs. 5-41, 5-43 and Table 5-2, we have

$$\psi_1(x) = \frac{\alpha^{3/4}}{\pi^{1/4}}xe^{-\alpha x^2/2}$$

and

$$\psi_2(x) = \left[\frac{\alpha}{64\pi}\right]^{1/4}(2\alpha x^2 - 2)e^{-\alpha x^2/2}$$

For n = 1, the Schrödinger equation becomes

$$\frac{d^2}{dx^2}(xe^{-\alpha x^2/2}) + \frac{2\mu}{\hbar^2}\left[\frac{3}{2}h\nu - \frac{1}{2}kx^2\right]xe^{-\alpha x^2/2} \stackrel{?}{=} 0$$

$$(-3\alpha x + \alpha^2 x^3) + \frac{2\mu}{\hbar^2}\left[\frac{3}{2}h\nu - \frac{1}{2}kx^2\right]x \overset{?}{=} 0$$

$$(\alpha^2 x^3 - 3\alpha x) + \frac{3}{\hbar}(\mu k)^{1/2}x - \frac{k\mu}{\hbar^2}x^3 \overset{?}{=} 0$$

and this does equal zero because $\alpha = (\mu k)^{1/2}/\hbar$.

For $n = 2$, we have

$$(6\alpha - 12\alpha^2 x^2 + 2\alpha^3 x^4)e^{-\alpha x^2/2} + \frac{2\mu}{\hbar^2}\left[\frac{5}{2}h\nu - \frac{1}{2}kx^2\right](4\alpha x^2 - 2)e^{-\alpha x^2/2} \overset{?}{=} 0$$

$$(6\alpha - 12\alpha^2 x^2 + 2\alpha^3 x^4) + (5\alpha - \alpha^2 x^2)(4\alpha x^2 - 2) \overset{?}{=} 0$$

$$10\alpha - 22\alpha^2 x^2 + 4\alpha^3 x^4 + (20\alpha^2 x^2 - 4\alpha^3 x^4 - 10\alpha + 2\alpha^2 x^2) = 0$$

Thus $\psi_2(x)$ satisfies the Schrödinger equation for a harmonic oscillator.

**5-12.** Consider the product of two even functions

$$P(x) = f(x)g(x)$$

Then

$$P(-x) = f(-x)g(-x) = f(x)g(x) = P(x)$$

and so $P(x)$ is an even function of $x$.

If $P(x)$ is the product of two odd functions of $x$, then

$$P(x) = f(x)g(x)$$

and

$$P(-x) = f(-x)g(-x) = (-)^2 f(x)g(x) = P(x)$$

and so $P(x)$ is an even function of $x$.

Lastly, if $P(x)$ is a product of an even function and an odd function of $x$, then

$$P(x) = f(x)g(x)$$

and

$$P(-x) = f(-x)g(-x) = -f(x)g(x)$$

and so $P(x)$ is an odd function of $x$.

**5-13.** If $f(x)$ is an even function of $x$, then it can be represented by a power series of the form

$$f(x) = f_0 + f_2x^2 + f_4x^4 + \cdots$$

The derivative of $f(x)$ is

$$f'(x) = 2f_2x + 4f_4x^3 + \cdots$$

which is an odd function of $x$.

Similarly, if $f(x)$ is an odd function of $x$, then it can be represented by a power series of the form

$$f(x) = f_1x + f_3x^3 + f_5x^5 + \cdots$$

The derivative of $f(x)$ is

$$f'(x) = f_1 + 3f_3x^2 + 5f_5x^4 + \cdots$$

which is an even function of $x$.

**5-14.** From Eqs. 5-41 and 5-43, we have

$$\psi_0(x) = N_0e^{-\alpha x^2/2}$$

$$\psi_1(x) = N_1(2\alpha^{1/2}x)e^{-\alpha x^2/2}$$

$$\psi_2(x) = N_2(4\alpha x^2 - 2)e^{-\alpha x^2/2}$$

$$\psi_3(x) = N_3(8\alpha^{3/2}x^3 - 12\alpha^{1/2}x)e^{-\alpha x^2/2}$$

There are five integrals that we must evaluate. Three of them are

$$\int_{-\infty}^{\infty} dx\,\psi_0(x)\psi_1(x) = \int_{-\infty}^{\infty} dx\,\psi_0(x)\psi_3(x)$$

$$= \int_{-\infty}^{\infty} dx\,\psi_1(x)\psi_2(x) = 0$$

because in each case the integrand is an odd function of $x$. This leaves two integrals:

$$\int_{-\infty}^{\infty} dx\,\psi_0(x)\psi_2(x) = 2N_0N_2\int_0^{\infty} dx\,(4\alpha x^2 - 2)e^{-\alpha x^2}$$

$$= 2N_0N_2\left[4\alpha\left[\frac{1}{4\alpha}\right]\left[\frac{\pi}{\alpha}\right]^{1/2} - 2\left[\frac{\pi}{4\alpha}\right]^{1/2}\right]$$

$$= 0$$

$$\int_{-\infty}^{\infty} dx\,\psi_1(x)\psi_3(x)$$

$$= 4\alpha^{1/2}N_1N_3\int_0^{\infty} dx\,(8\alpha^{3/2}x^4 - 12\alpha^{1/2}x^2)e^{-\alpha x^2}$$

$$= 4\alpha^{1/2}N_1N_3\left[8\alpha^{3/2}\left[\frac{3}{8\alpha^2}\right]\left[\frac{\pi}{\alpha}\right]^{1/2} - 12\alpha^{1/2}\left[\frac{1}{4\alpha}\right]\left[\frac{\pi}{\alpha}\right]^{1/2}\right]$$

$$= 0$$

**5-15.** Because $x^{2n}e^{-\alpha x^2}$ is an even function of $x$ ($n$ is an integer), we have that

$$I_n(a) = \int_{-\infty}^{\infty} dx\ x^{2n}e^{-ax^2} = 2\int_0^{\infty} dx\ x^{2n}e^{-ax^2}$$

For $n = 0$, we write

$$I_0^2(a) = 4\int_0^{\infty} dx e^{-ax^2}\int_0^{\infty} dy e^{-ay^2}$$

$$= 4\int_0^{\infty}\int_0^{\infty} dx\ dy\ e^{-a(x^2+y^2)}$$

Let $dxdy = rdrd\theta$ and $x^2 + y^2 = r^2$ and realize that we are integrating over the entire first quadrant ($0 \le r < \infty$, $0 \le \theta \le \pi/2$) to write

$$I_0^2(a) = 4\int_0^{\pi/2} d\theta\ \int_0^{\infty} dr\ re^{-ar^2}$$

$$= 2\pi\left[\frac{1}{2a}\right] = \frac{\pi}{a}$$

or that

$$I_0(a) = \int_{-\infty}^{\infty} dx e^{-ax^2} = \left[\frac{\pi}{a}\right]^{1/2}$$

Now differentiate $I_0(a)$ with respect to $a$

$$\frac{dI_0}{da} = -\int_{-\infty}^{\infty} dx\ x^2e^{-ax^2} = -I_1(a)$$

$$\frac{d^2I_0}{da^2} = \int_{-\infty}^{\infty} dx\ x^4e^{-ax^2} = I_2(a)$$

and so forth to obtain

$$\frac{d^n I_0}{da^n} = (-1)^n \int_{-\infty}^{\infty} dx\ x^{2n} e^{-ax^2} = (-1)^n I_n(a)$$

**5-16.** From Eqs. 5-41 and 5-43

$$\psi_2(x) = \left[\frac{\alpha}{64\pi}\right]^{1/4} (4\alpha x^2 - 2) e^{-\alpha x^2/2}$$

Thus

$$\langle x^2 \rangle = \left[\frac{\alpha}{64\pi}\right]^{1/2} \int_{-\infty}^{\infty} dx\, (4\alpha x^2 - 2)^2 x^2 e^{-\alpha x^2}$$

$$= \left[\frac{\alpha}{64\pi}\right]^{1/2} \int_{-\infty}^{\infty} dx\, (16\alpha^2 x^6 - 16\alpha x^4 + 4x^2) e^{-\alpha x^2}$$

$$= \left[\frac{\alpha}{64\pi}\right]^{1/2} \left[16\alpha^2 \left[\frac{30}{16\alpha^3}\right] - 16\alpha\left[\frac{6}{8\alpha^2}\right] + 4\left[\frac{2}{4\alpha}\right]\right] \left[\frac{\pi}{\alpha}\right]^{1/2}$$

$$= \frac{5}{2\alpha} = \frac{5}{2}\frac{\hbar}{(\mu k)^{1/2}} = \frac{5\hbar}{(\mu k)^{1/2}}$$

**5-17.** From Eqs. 5-41 and 5-43,

$$\psi_2(x) = \left[\frac{\alpha}{64\pi}\right]^{1/4} (4\alpha x^2 - 2) e^{-\alpha x^2/2}$$

Recall that

$$\hat{P}_x^2 = -\hbar^2 \frac{d^2}{dx^2}$$

and so

$$\hat{P}_x^2 \psi_2(x) = \left[\frac{\alpha}{64\pi}\right]^{1/4} (10\alpha - 22\alpha^2 x^2 + 4\alpha^3 x^4) e^{-\alpha x^2/2}$$

Therefore

$$\langle p^2 \rangle = \int_{-\infty}^{\infty} dx\, \psi_2(x) \hat{P}_x{}^2 \psi_2(x)$$

$$= -\hbar^2 \left[\frac{\alpha}{64\pi}\right]^{1/2} \int_{-\infty}^{\infty} dx\, (4\alpha x^2 - 2)(10\alpha - 22\alpha^2 x^2 + 4\alpha^3 x^4) e^{-\alpha x^2}$$

$$= -\hbar^2\left[\frac{\alpha}{64\pi}\right]^{1/2}\int_{-\infty}^{\infty} dx\,(84\alpha^2x^2 - 96\alpha^3x^4 + 16\alpha^4x^6 - 20\alpha)e^{-\alpha x^2}$$

$$= -\hbar^2\left[\frac{\alpha}{64\pi}\right]^{1/2}\left[84\alpha^2\left[\frac{2}{4\alpha}\right] - 96\alpha^3\left[\frac{6}{8\alpha^2}\right] + 16\alpha^4\left[\frac{30}{16\alpha^3}\right] - 20\alpha\right]\left[\frac{\pi}{\alpha}\right]^{1/2}$$

$$= \frac{20\alpha\hbar^2}{8} = \frac{5}{2}\alpha\hbar^2 = \frac{5}{2}\,\frac{\hbar}{(\mu k)^{1/2}}$$

**5-18.** This problem is similar to Example 5-10. We use the formula

$$A_{rms} = \left[\frac{\hbar}{4\pi c\bar{\nu}_{obs}\mu}\right]^{1/2}$$

Thus, for $H_2$

$$A_{rms} = \left[\frac{1.055 \times 10^{-34}\ \text{J·s}}{4\pi(2.998 \times 10^{10}\ \text{cm·s}^{-1}) \times (4.33 \times 10^3\ \text{cm}^{-1})((1/2)\text{amu}) \times (1.6607 \times 10^{-27}\ \text{kg·amu}^{-1})}\right]^{1/2}$$

$$= 8.82 \times 10^{-12}\ \text{m} = 8.82\ \text{pm}$$

For $^{35}Cl^{35}Cl$

$$A_{rms} = \left[\frac{1.055 \times 10^{-34}\ \text{J·s}}{4\pi(2.998 \times 10^{10}\ \text{cm·s}^{-1}) \times (560\ \text{cm}^{-1})((35/2)\text{amu}) \times (1.6607 \times 10^{-27}\ \text{kg·amu}^{-1})}\right]^{1/2}$$

$$= 4.15 \times 10^{-12}\ \text{m} = 4.15\ \text{pm}$$

For $^{14}N^{14}N$

$$A_{rms} = \left[\frac{1.055 \times 10^{-34}\ \text{J·s}}{4\pi(2.998 \times 10^{10}\ \text{cm·s}^{-1}) \times (2.35 \times 10^3\ \text{cm})(7\ \text{amu}) \times (1.6607 \times 10^{-27}\ \text{kg·amu}^{-1})}\right]^{1/2}$$

$$= 3.20 \times 10^{-12}\ \text{m} = 3.20\ \text{pm}$$

**5-19.** Eq. 5-69 is

$$\frac{d^2\psi}{dx^2} + (\lambda - \alpha^2x^2)\psi = 0$$

If we let $x = \beta\xi$, then this equation becomes

$$\frac{d^2\psi}{\beta^2 d\xi^2} + (\lambda - \alpha^2\beta^2\xi^2)\psi = 0$$

Multiply through by $\beta^2$

$$\frac{d^2\psi}{d\xi^2} + (\lambda\beta^2 - \alpha^2\beta^4\xi^2)\psi = 0$$

We choose $\beta$ so that the coefficient of $\xi^2$ is unity, or $\alpha^2\beta^4 = 1$, or

$$\beta = 1/(\alpha^{1/2})$$

Thus the substitution

$$\xi = \alpha^{1/2}x$$

converts Eq. 5-69 to Eq. 5-71.

**5-20.** Eq. 5-71 is

$$\frac{d^2\psi}{d\xi^2} + \left[\frac{\lambda}{\alpha} - \xi^2\right]\psi = 0$$

Substitute $e^{\pm\xi^2}$ into this equation to get

$$(4\xi^2 \pm 2)e^{\pm\xi^2} + \left[\frac{\lambda}{\alpha} - \xi^2\right]e^{\pm\xi^2}$$

As $\xi^2$ becomes large (asymptotic solution), this expression becomes

$$(4\xi^2 - \xi^2)e^{\pm\xi^2} \neq 0$$

If we had used $e^{\pm\beta\xi^2}$ instead of $e^{\pm\xi^2}$, we would have obtained

$$(4\beta^2 - 1)\xi^2 e^{\pm\beta\xi^2} = 0$$

or

$$\beta = \pm 1/2$$

**5-21.** Start with Eq. 5-71

$$\frac{d^2\psi}{d\xi^2} + \left[\frac{\lambda}{\alpha} - \xi^2\right]\psi = 0$$

Substitute $\psi(\xi) = H_n(\xi)e^{-\xi^2/2}$ into this equation to obtain

$$\left[\frac{d^2H}{d\xi^2} - 2\xi\frac{dH}{d\xi} + (\xi^2 - 1)H\right] + \left[\frac{\lambda}{\alpha} - \xi^2\right]H = 0$$

where we have cancelled a factor of $e^{-\xi^2/2}$.

Collecting terms, we have

$$\frac{d^2H}{d\xi^2} - 2\xi\frac{dH}{d\xi} + \left[\frac{\lambda}{\alpha} - 1\right]H = 0 \qquad (5\text{-}76)$$

**5-22.** Eq. 5-76 with $(\lambda/\alpha) - 1 = 2n$ is

$$\frac{d^2H_n}{d\xi^2} - 2\xi\frac{dH_n}{d\xi} + 2nH_n = 0$$

If $n = 0$, $H_0(\xi) = 1$ clearly satisfies Hermite's equation.

For $n = 1$, we have

$$0 - 2\xi(2) + 2(2\xi) = 0$$

For $n = 2$,

$$8 - 2\xi(8\xi) + 4(4\xi^2 - 2) = 0$$

For $n = 3$,

$$(48\xi) - 2\xi(24\xi^2 - 12) + 6(8\xi^3 - 12\xi) = 0$$

**5-23.** Hermite's differential equation is (Eq. 5-76)

$$\frac{d^2H_n}{d\xi^2} - 2\xi\frac{dH_n}{d\xi} + \left[\frac{\lambda}{\alpha} - 1\right]H_n(\xi) = 0$$

Substitution of $H(\xi) = a_0 + a_1\xi + a_2\xi^2 + a_3\xi^3 + \cdots$ into this equation gives

$$[2a_2 + 2\cdot3a_3\xi + 3\cdot4a_4\xi^2 + \cdots]$$
$$- 2\xi[a_1 + 2a_2\xi + 3a_3\xi^2 + \cdots]$$
$$+ [\lambda/\alpha - 1][a_0 + a_1\xi + a_2\xi^2 + \cdots] = 0$$

Collect like powers of $\xi$ to obtain

$$[2a_2 + (\lambda/\alpha - 1)a_0] + [2\cdot3a_3 + (\lambda/\alpha - 1 - 2)a_1]\xi$$
$$+ [3\cdot4a_4 + (\lambda/\alpha - 1 - 2\cdot2)a_2]\xi^2$$
$$+ [4\cdot5a_5 + (\lambda/\alpha - 1 - 2\cdot3)a_3]\xi^3 + \cdots$$
$$= 0$$

Now equate each power of $\xi$ to zero to obtain

$$1\cdot 2a_2 + (\lambda/\alpha - 1)a_0 = 0$$

$$2\cdot 3a_3 + (\lambda/\alpha - 1 - 2)a_1 = 0$$

$$3\cdot 4a_4 + (\lambda/\alpha - 1 - 2\cdot 2)a_2 = 0$$

$$4\cdot 5a_5 + (\lambda/\alpha - 1 - 2\cdot 3)a_3 = 0$$

and so on. The general term (by inspection) is

$$(n + 1)(n + 2)a_{n+2} + (\lambda/\alpha - 1 - 2n)a_n = 0$$

**5-24.** From Problem 5-23 we have

$$\frac{a_{n+2}}{a_n} = \frac{(\lambda/\alpha - 1 - 2n)}{(n + 1)(n + 2)}$$

For large values of n, we can neglect$(\lambda/\alpha) - 1$ and get

$$\left|\frac{a_{n+2}}{a_n}\right| \to \frac{2n}{n^2} = \frac{2}{n}$$

**5-25.** From Eq. 5-80 we have

$$(n + 1)(n + 2)a_{n+2} + (\lambda/\alpha - 1 - 2n)a_n = 0$$
$$n = 0,1,2,\cdots$$

If

$$\lambda/\alpha - 1 = 2n$$

for a given value of n, then Eq. 5-80 becomes

$$a_{n+2} = 0$$

But if $a_{n+2} = 0$ for a given value of n, then

$$a_{n+4} = a_{n+6} = a_{n+8} = \cdots = 0$$

ad so all the coefficients of the powers of $\xi$ beyond $\xi^n$ are zero. Thus the functions $H(\xi)$ are polynomials in $\xi$.

**5-26.** Set $(\lambda/\alpha) - 1 = 2n$ in Eqs. 5-82 and 5-83 to obtain

$$a_2 = -\frac{2na_0}{2!}$$

$$a_4 = \frac{(2n)(2n - 4)a_0}{4!}$$

Using Eq. 5-80 with n = 4, we can see that

$$a_6 = -\frac{(\lambda/\alpha - 1 - 8)a_4}{5\cdot 6} = -\frac{(2n - 8)a_4}{5\cdot 6}$$

$$= -\frac{(2n)(2n - 4)(2n - 8)a_0}{6!}$$

where we have used the above expression for $a_4$.

Therefore, we have for $H_n(\xi)$

$$H_n(\xi) = a_0\left[1 - \frac{2n}{2!}\xi^2 + \frac{2^2 n(n-2)}{4!}\xi^4 - \frac{2^3 n(n-2)(n-4)}{6!}\xi^6 + \cdots\right]$$

Starting with Eqs. 5-84 and 5-85, we have

$$a_3 = -\frac{(2n - 2)}{2\cdot 3}a_1$$

$$a_5 = \frac{(2n - 2)(2n - 6)}{5!}a_1$$

Using Eq. 5-80 with n = 5,

$$a_7 = -\frac{(2n - 10)}{6\cdot 7}a_5$$

$$= -\frac{(2n - 2)(2n - 6)(2n - 10)}{7!}a_1$$

Thus we have for $H(\xi)$

$$H_n(\xi) = a_1\left[\xi - \frac{2(n - 1)}{3!}\xi^3 + \frac{2^2(n - 1)(n - 3)}{5!}\xi^5 - \frac{2^3(n - 1)(n - 3)(n - 5)}{7!}\xi^7 + \cdots\right]$$

**5-27.** We shall verify the formulas in Problem 5-27 for n = 0, 1, and 2.

$n = 0$: $\quad H_0(\xi) = 1,$

$\qquad\qquad H_1(\xi) = 2\xi$

$0 = 2\xi(1) - 2\xi = 0$

$2\xi - 2\xi(1) + 0 = 0$

$0 = 0$

$n = 1$:  $H_0(\xi) = 1,$

$H_1(\xi) = 2\xi,$

$H_2(\xi) = 4\xi^2 - 2$

$2 = 2\xi(2\xi) - 4\xi^2 + 2 = 2$

$(4\xi^2 - 2) - 2\xi(2\xi) + 2(1) = 0$

$2 = 2(1)$

$n = 2$:  $H_1(\xi) = 2\xi,$

$H_2(\xi) = 4\xi^2 - 2,$

$H_3(\xi) = 8\xi^3 - 12\xi$

$8\xi = 2\xi(4\xi^2 - 2) - (8\xi^3 - 12\xi) = 8\xi$

$(8\xi^3 - 12\xi) - 2\xi(4\xi^2 - 2) + 4(2\xi) = 0$

$8\xi = 4(2\xi) = 8\xi$

**5-28.**

$$\langle p\rangle = \int_{-\infty}^{\infty} \psi_n(x)\hat{P}_x\psi_n(x)\,dx$$

$$= -i\hbar\int_{-\infty}^{\infty} \psi_n(x)\left[\frac{d}{dx}\right]\psi_n(x)\,dx$$

Because $\psi_n(x)$ is an even function of x if n is even and an odd function of x if n is odd,

$$\frac{d\psi_n}{dx} = \text{an even function of x if n is odd}$$

$$= \text{an odd function of x if n is even}$$

(see Problem 5-13). Consequently the integrand in $\langle p\rangle$ will be an odd function of x for any value of n and so $\langle p\rangle = 0$.

$$\langle p\rangle = \int_{-\infty}^{\infty} (\text{odd function of x})\,dx = 0$$

Now consider

$$\langle p^2\rangle = \int_{-\infty}^{\infty} \psi_n(x)\hat{P}_x{}^2\psi_n(x)\,dx$$

$$= -\hbar^2\int_{-\infty}^{\infty} \psi_n(x)\left[\frac{d^2}{dx^2}\right]\psi_n(x)\,dx$$

First let $\xi = \alpha^{1/2}x$ and write

$$\langle p^2\rangle = -\alpha^{1/2}\hbar^2\int_{-\infty}^{\infty}\psi_n(\xi)\left[\frac{d^2}{d\xi^2}\right]\psi_n(\xi)d\xi$$

Now evaluate $\psi_n''(\xi)$

$$\frac{d\psi_n}{d\xi} = N_n\left[\frac{d}{d\xi}(H_n(\xi)e^{-\xi^2/2})\right]$$

$$= N_n\left[H_n'(\xi)e^{-\xi^2/2} - 2\xi H_n(\xi)e^{-\xi^2/2}\right]$$

$$\frac{d^2\psi_n}{d\xi^2} = N_n\left[H_n''(\xi)e^{-\xi^2/2} - 2\xi H_n'(\xi)e^{-\xi^2/2} + (\xi^2 - 1)H_n(\xi)e^{-\xi^2/2}\right]$$

Therefore

$$\langle p^2\rangle = -\alpha^{1/2}\hbar^2N_n^2\int_{-\infty}^{\infty}d\xi e^{-\xi^2}H_n(\xi)[H_n''(\xi) - 2\xi H_n'(\xi) + (\xi^2-1)H_n(\xi)]$$

From Hermite's differential equation, we see that

$$H_n''(\xi) - 2\xi H_n'(\xi) = -2nH_n(\xi)$$

Substitute this result into $\langle p^2\rangle$ to get

$$\langle p^2\rangle = -\alpha^{1/2}\hbar^2N_n^2\int_{-\infty}^{\infty}d\xi e^{-\xi^2}H_n(\xi)[(\xi^2 - 1 - 2n)]H_n(\xi)$$

$$= -\alpha\hbar^2[\langle\xi^2\rangle - (2n + 1)]$$

But $\langle\xi^2\rangle = n + 1/2$ from page 183, and so

$$\langle p^2\rangle = -\alpha\hbar^2[n + 1/2 - 2(n + 1/2)]$$

$$= (n + 1/2)\hbar^2\alpha = (n + 1/2)\hbar(\mu k)^{1/2}$$

**5-29.** The Hamiltonian operator of a harmonic oscillator is

$$\hat{H}(x) = -\frac{\hbar^2}{2\mu}\frac{d^2}{dx^2} + \frac{1}{2}kx^2$$

If we replace $x$ by $-x$, $\hat{H}(x)$ does not change, or

$$\hat{H}(x) = \hat{H}(-x)$$

Now consider the Schrödinger equation of a harmonic oscillator

$$\hat{H}(x)\psi_n(x) = E_n\psi_n(x)$$

Replace x by −x and use the fact that $\hat{H}(x) = \hat{H}(-x)$ to obtain

$$\hat{H}\psi_n(-x) = E_n\psi_n(-x)$$

Thus, we see that $\psi_n(-x)$ is also an eigenfunction of $\hat{H}(x)$ belonging to the same eigenvalue $E_n$. Since the system is nondegenerate, $\psi_n(x)$ and $\psi_n(-x)$ differ by only a multiplicative constant. We can write this condition as

$$\psi_n(x) = c\psi_n(-x)$$

But $\psi_n(-x) = c\psi_n(x)$, and so

$$\psi_n(x) = c^2\psi_n(-x)$$

which implies that $c^2 = \pm 1$. Thus

$$\psi_n(x) = \pm\psi_n(-x)$$

We define a reflection operator $\hat{R}$ by

$$\hat{R}u(x) = u(-x)$$

The operator $\hat{R}$ is linear because

$$\begin{aligned}\hat{R}(c_1u_1(x) + c_2u_2(x)) &= c_1u_1(-x) + c_2u_2(-x)\\ &= c_1\hat{R}u_1(x) + c_2\hat{R}u_2(x)\end{aligned}$$

Because $\hat{H}(x) = \hat{H}(-x)$

$$\hat{H}(x)\hat{R}\psi_n(x) = \hat{H}(x)\psi_n(-x) = E_n\psi_n(-x)$$

$$\hat{R}\hat{H}(x)\psi_n(x) = \hat{R}E_n\psi_n(x) = E_n\psi_n(-x)$$

Because $\hat{R}\psi_n(x) = \psi_n(-x) = \pm\psi_n(x)$, we see that the eigenvalues of $\hat{R}$ are ±1, and its eigenfunctions are $\psi_n(x)$. Furthermore, because $\hat{H}$ and $\hat{R}$ commute, they have mutual eigenfunctions (page 138).

**5-30.** Eq. 5-71 is

$$\frac{d^2\psi}{d\xi^2} + \left[\frac{\lambda}{\alpha} - \xi^2\right]\psi = 0$$

Substitute

$$\psi(\xi) = a_0 + a_1\xi + a_2\xi^2 + a_3\xi^3 + \cdots$$

into the differential equation to obtain

$$\begin{aligned}&[2a_2 + 2\cdot 3a_3\xi + 3\cdot 4a_4\xi^2 + 4\cdot 5a_5\xi^3 + \cdots]\\ &\quad + (\lambda/\alpha)[a_0 + a_1\xi + a_2\xi^2 + a_3\xi^3 + \cdots]\\ &\quad - [a_0\xi^2 + a_1\xi^3 + \cdots] = 0\end{aligned}$$

Collect like powers of $\xi$

$$[2a_2 + (\lambda/2)a_0] + [2\cdot 3a_3 + (\lambda/\alpha)a_1]\xi$$
$$+ [3\cdot 4a_4 + (\lambda/\alpha)a_2 - a_0]\xi^2$$
$$+ [4\cdot 5a_5 + (\lambda/\alpha)a_3 - a_1]\xi^3 + \cdots = 0$$

Note that setting the coefficients of $\xi^n$, $n \geq 2$, equal to zero leads to three-term recursion formulas.

**5-31.** The expression

$$dt = \frac{dx}{\omega[A^2 - x^2]^{1/2}}$$

gives the time that a classical harmonic oscillator spends between x and x + dx. The fraction of time that the oscillator spends between x and x + dx is

$$f(t)dt = \frac{dx}{2\pi\sqrt{A^2 - x^2}}$$

To show that this expression is normalized, we must integrate over one complete cycle of vibration, or twice the path −A to +A.

Thus

$$2\int_{-A}^{A} \frac{dx}{2\pi\sqrt{A^2 - x^2}} = \frac{1}{\pi}\left|_{-A}^{A} \sin^{-1}\frac{x}{A}\right| = 1$$

**5-32.** Substitute

$$y(x) = a_0 + a_1x + a_2x^2 + a_3x^3 + a_4x^4 + \cdots$$

into

$$y''(x) + k^2y(x) = 0$$

to obtain

$$[2a_2 + 2\cdot 3a_3x + 3\cdot 4a_4x^2$$
$$+ 4\cdot 5a_5x^3 + 5\cdot 6a_6x^4 + \cdots]$$
$$+ k^2[a_0 + a_1x + a_2x^2 + a_3x^3 + a_4x^4 + \cdots]$$
$$= 0$$

Collect like powers of x

$$(2a_2 + k^2a_0) + (2\cdot 3a_3 + k^2a_1)x + (3\cdot 4a_4 + k^2a_2)x^2 + (4\cdot 5a_5 + k^2a_3)x^3 + \cdots = 0$$

Set each coefficient equal to zero and obtain the general result

$$n(n-1)a_n + k^2a_{n-2} = 0 \qquad n \geq 2$$

Using this recursion relation for n = 2,3,..., we find

$$a_2 = -\frac{k^2}{2}a_0 \qquad a_3 = -\frac{k^2}{3\cdot 2}a_1$$

$$a_4 = -\frac{k^2}{4\cdot 3}a_2 = \frac{k^4}{4!}a_0 \qquad a_5 = -\frac{k^2}{5\cdot 4}a_3 = \frac{k^4}{5!}a_1$$

$$a_6 = -\frac{k^2}{6\cdot 5}a_4 = -\frac{k^6}{6!}a_0 \qquad a_7 = -\frac{k^2}{7\cdot 6}a_5 = -\frac{k^6}{7!}a_1$$

and so on. Thus

$$y(x) = a_0\left[1 - \frac{k^2x^2}{2!} + \frac{k^4x^4}{4!} - \frac{k^6x^6}{6!} + \cdots\right] + a_1\left[x - \frac{k^2x^3}{3!} + \frac{k^4x^5}{5!} - \frac{k^6x^7}{7!} + \cdots\right]$$

Let $a_1' = a_1/k$ and obtain

$$y(x) = a_0C(kx) + a_1'S(kx)$$

where

$$C(\xi) = 1 - \frac{\xi^2}{2!} + \frac{\xi^4}{4!} - \frac{\xi^6}{6!} + \cdots$$

and

$$S(\xi) = \xi - \frac{\xi^3}{3!} + \frac{\xi^5}{5!} - \frac{\xi^7}{7!} + \cdots$$

If we differentiate $S(\xi)$ term by term, then we find

$$\frac{dS}{d\xi} = 1 - \frac{\xi^2}{2!} + \frac{\xi^4}{4!} - \frac{\xi^6}{6!} + \cdots = C(\xi)$$

and if we differentiate $C(\xi)$ term by term

$$\frac{dC}{d\xi} = -\xi + \frac{\xi^3}{3!} - \frac{\xi^5}{5!} + \cdots = -S(\xi)$$

The power series for $C(\xi) + iS(\xi)$ is

$$1 + i\xi - \frac{\xi^2}{2!} - i\frac{\xi^3}{3!} + \frac{\xi^4}{4!} + i\frac{\xi^5}{5!} - \frac{\xi^6}{6!} - i\frac{\xi^7}{7!} + \cdots$$

Introducing $i^2 = -1$, we can rewrite this power series as

$$1 + i\xi + \frac{i^2\xi^2}{2!} + \frac{i^3\xi^3}{3!} + \frac{i^4\xi^4}{4!} + \frac{i^5\xi^5}{5!} + \cdots$$

which is the power series expansion of $e^{i\xi}$.

To prove that $C^2(\xi) + S^2(\xi) = 1$, we write $C^2(\xi)$ and $S^2(\xi)$ to $O(\xi^8)$

$$C^2(\xi) = 1 - \xi^2 + \left[\frac{1}{12} + \frac{1}{4}\right]\xi^4 - \left[\frac{1}{4!} + \frac{2}{6!}\right]\xi^6 + O(\xi^8)$$

$$S^2(\xi) = \xi^2 - \frac{\xi^4}{3} + \left[\frac{2}{5!} + \frac{1}{(3!)^2}\right]\xi^6 + O(\xi^8)$$

$$C^2(\xi) + S^2(\xi) = 1 + O(\xi^8)$$

To prove that $S(2\xi) = 2S(\xi)C(\xi)$, we first multiply $S(\xi)$ by $C(\xi)$ to $O(\xi^8)$

$$S(\xi)C(\xi) = \xi - \left[\frac{1}{3!} + \frac{1}{2!}\right]\xi^3 + \left[\frac{1}{5!} + \frac{1}{2!3!} + \frac{1}{4!}\right]\xi^5 - \left[\frac{1}{7!} + \frac{1}{2!5!} + \frac{1}{4!3!} + \frac{1}{6!}\right]\xi^7 + \cdots$$

$$= \xi - \frac{4}{3!}\xi^3 + \frac{16}{5!}\xi^5 - \frac{64}{7!}\xi^7 + \cdots$$

and

$$S(2\xi) = 2\xi - \frac{8\xi^3}{3!} + \frac{32\xi^5}{5!} - \frac{128\xi^7}{7!} + \cdots$$

and so we see that

$$S(2\xi) = 2S(\xi)C(\xi)$$

at least to $O(\xi^8)$.

To prove that $S^2(\xi) = (1/2)[1 - C(\xi)]$, we need $C(2\xi)$

$$C(2\xi) = 1 - \frac{4\xi^2}{2!} + \frac{16\xi^4}{4!} - \frac{64\xi^6}{6!} + O(\xi^8)$$

$$\frac{1}{2}[1 - C(2\xi)] = \xi^2 - \frac{1}{3}\xi^4 + \frac{2}{45}\xi^6 + O(\xi^8)$$

Comparing this to $S^2(\xi)$, which was calculated above, we see that

$$S^2(\xi) = (1/2)[1 - C(2\xi)]$$

at least to $O(\xi^8)$.

**5-33.** The normal distribution is

$$\phi(z)dz = (2\pi\sigma^2)^{-1/2} \exp\left[-\frac{(z - \langle z\rangle)^2}{2\sigma^2}\right]dz$$

Note that

$$\langle z\rangle = (2\pi\sigma^2)^{-1/2}\int_{-\infty}^{\infty} z \exp[-(z - \langle z\rangle)^2/2\sigma^2]dz$$

$$= (2\pi\sigma^2)^{-1/2}\int_{-\infty}^{\infty}(y + \langle z\rangle)\exp(-y^2/2\sigma^2)dy = \langle z\rangle$$

and that

$$\langle z^2\rangle = (2\pi\sigma^2)^{-1/2}\int_{-\infty}^{\infty} z^2\exp[-(z - \langle z\rangle)^2/2\sigma^2]dz$$

$$= (2\pi\sigma^2)^{-1/2}\int_{-\infty}^{\infty}(y^2 + 2y\langle z\rangle + \langle z\rangle^2)e^{-y^2/2\sigma^2}dy$$

$$= \sigma^2 + 0 + \langle z\rangle^2$$

Introduce the standardized variable

$$x = \frac{z - \langle z\rangle}{\sigma}$$

to get

$$\phi(x)dx = (2\pi)^{-1/2}e^{-x^2/2}dx$$

To derive this result, we used the fact that $dz = \sigma dx$. Note that

$$\langle x\rangle = \left\langle\frac{z - \langle z\rangle}{\sigma}\right\rangle = \frac{\langle z\rangle - \langle z\rangle}{\sigma} = 0$$

and that

$$\langle x^2\rangle = \left\langle \frac{z^2 - 2z\langle z\rangle + \langle z\rangle^2}{\sigma^2} \right\rangle = \frac{\langle z^2\rangle - \langle z\rangle^2}{\sigma^2}$$

$$= \frac{\sigma^2 + \langle z\rangle^2 - \langle z\rangle^2}{\sigma^2} = 1$$

Multiply

$$f(x) = e^{-x^2/2}[a_0He_0(x) + a_1He_1(x) + a_2He_2(x) + \cdots]$$

by $He_j(x)$ and integrate to get

$$\int_{-\infty}^{\infty} He_j(x)f(x)dx = a_j\int_{-\infty}^{\infty} e^{-x^2/2}He_j{}^2(x)dx = (2\pi)^{1/2}j!$$

or

$$a_j = \frac{1}{(2\pi)^{1/2}j!}\int_{-\infty}^{\infty} He_j(x)f(x)dx$$

To evaluate the first few $a_j$, we must generate the first few $He_j(x)$.

$$He_0(x) = 1$$

$$He_1(x) = (-1)e^{x^2/2}\frac{d}{dx}e^{-x^2/2} = x$$

$$He_2(x) = e^{x^2/2}\frac{d^2}{dx^2}e^{-x^2/2} = x^2 - 1$$

Thus

$$a_0 = (2\pi)^{-1/2}\int_{-\infty}^{\infty} He_0(x)f(x)dx = (2\pi)^{-1/2}\int_{-\infty}^{\infty} f(x)dx$$

$$= (2\pi)^{-1/2}$$

since $f(x)$ is a probability density.

$$a_1 = (2\pi)^{-1/2}\int_{-\infty}^{\infty} He_1(x)f(x)dx$$

$$= (2\pi)^{-1/2}\int_{-\infty}^{\infty} x\ f(x)dx = 0$$

because $\langle x\rangle = 0$.

$$a_2 = (2\pi)^{-1/2}\int_{-\infty}^{\infty} He_2(x)f(x)dx$$

$$= (2\pi)^{-1/2}\int_{-\infty}^{\infty}(x^2 - 1)f(x)dx = 0$$

because $\langle x^2 \rangle = 1$.

The equation for f(x) becomes

$$f(x) = (2\pi)^{-1/2}e^{-x^2/2} + e^{-x^2/2}\sum_{n=3}^{\infty} a_n He_n(x)$$

## CHAPTER SIX

## THREE-DIMENSIONAL SYSTEMS

**6-1.** The normalization condition is

$$\int_0^a\int_0^b\int_0^c \psi^*(x,y,z)\psi(x,y,z)dxdydz = 1$$

Substitute Eq. 6-9 into this integral to obtain

$$\int_0^a A_x{}^2\sin^2\frac{n_x\pi x}{a}dx \int_0^b A_y{}^2\sin^2\frac{n_y\pi y}{b}dy \int_0^c A_z{}^2\sin^2\frac{n_z\pi z}{c}dz = 1$$

or

$$(A_xA_yA_z)^2\ (a/2)(b/2)(c/2) = 1$$

Thus the normalization constant is

$$A_xA_yA_z = \left[\frac{8}{abc}\right]^{1/2}$$

**6-2.** The average momentum of a particle in a three-dimensional box is given by

$$\langle\mathbf{p}\rangle = \int_0^a dx \int_0^b dy \int_0^c dz\ \psi^*(x,y,z)\hat{\mathbf{P}}\psi(x,y,z)$$

where

$$\hat{\mathbf{P}} = -i\hbar\left[\mathbf{i}\frac{\partial}{\partial x} + \mathbf{j}\frac{\partial}{\partial y} + \mathbf{k}\frac{\partial}{\partial z}\right]$$

If we substitute this operator into the expression for $\langle\mathbf{p}\rangle$, then we obtain

$$\langle\mathbf{p}\rangle = -i\hbar\mathbf{i}\int_0^a \sin\frac{n_x\pi x}{a}\ \frac{\partial}{\partial x}\left[\sin\frac{n_x\pi x}{a}\right]dx$$

$$\times \int_0^b \sin^2\frac{n_y\pi y}{b}dy \int_0^c \sin^2\frac{n_z\pi z}{c}dz$$

$$- i\hbar\mathbf{j}\int_0^a \sin^2\frac{n_x\pi x}{a}dx \int_0^b \sin\frac{n_y\pi y}{b}\ \frac{\partial}{\partial y}\left[\sin\frac{n_y\pi y}{b}\right]dy$$

$$\times \int_0^c \sin^2\frac{n_z\pi z}{c}dz$$

$$- i\hbar\mathbf{k}\int_0^a \sin^2\frac{n_x\pi x}{a}dx \int_0^b \sin^2\frac{n_y\pi y}{b}dy$$

$$\times \int_0^c \sin\frac{n_z\pi z}{c}\ \frac{\partial}{\partial z}\left[\sin\frac{n_z\pi z}{c}\right]dz$$

$$= 0$$

Each of these three integrals has a factor like

$$\int_0^a \sin\frac{n_x\pi x}{a}\cos\frac{n_x\pi x}{a}dx$$

that results from taking a derivative. This integral is equal to zero, and so we have

$$\langle \mathbf{p}\rangle = 0$$

**6-3.** The ground state wave function for a particle in a cube of length a is

$$\psi = \left[\frac{8}{a^3}\right]^{1/2}\sin\frac{\pi x}{a}\ \sin\frac{\pi y}{a}\ \sin\frac{\pi z}{a}$$

The probability that the particle is in the subcube given in the problem is

$$\iiint_{a/4}^{3a/4} \psi^*\psi dx dy dz$$

$$= \left[\frac{8}{a^3}\right]\int_{a/4}^{3a/4}\sin^2\frac{\pi x}{a}dx \int_{a/4}^{3a/4}\sin^2\frac{\pi y}{a}dy \int_{a/4}^{3a/4}\sin^2\frac{\pi z}{a}dz$$

Each integral is equal to $a((1/4) + (1/2\pi))$, and so

$$\text{probability} = \left[\frac{1}{2} + \frac{1}{\pi}\right]^3 = .548$$

**6-4.** This problem is done in three dimensions on pages 196 and 197 of the text. The analogs of the equations given there are

$$-\frac{\hbar^2}{2m}\left[\frac{\partial^2}{\partial x^2} + \frac{\partial^2}{\partial y^2}\right]\psi(x,y) = E\psi(x,y) \qquad \begin{matrix} 0 \le x \le a \\ 0 \le y \le b \end{matrix} \tag{6-1}$$

$$\begin{aligned} \psi(0,y) &= \psi(a,y) = 0 \quad \text{for all } y \\ \psi(x,0) &= \psi(x,b) = 0 \quad \text{for all } x \end{aligned} \tag{6-2}$$

$$\psi(x,y) = X(x)Y(y) \tag{6-3}$$

$$-\frac{\hbar^2}{2m}\frac{X''}{X} - \frac{\hbar^2}{2m}\frac{Y''}{Y} = E \tag{6-4}$$

$$E_x + E_y = E \tag{6-5}$$

$$\begin{aligned} -\frac{\hbar^2}{2m}\frac{X''}{X} &= E_x \\ -\frac{\hbar^2}{2m}\frac{Y''}{Y} &= E_y \end{aligned} \tag{6-6}$$

$$\begin{aligned} X(0) &= X(a) = 0 \\ Y(0) &= Y(b) = 0 \end{aligned} \tag{6-7}$$

$$\begin{aligned} X(x) &= A_x \sin\frac{n_x \pi x}{a} \qquad & n_x = 1,2,\ldots \\ Y(y) &= A_y \sin\frac{n_y \pi y}{b} \qquad & n_y = 1,2,\ldots \end{aligned} \tag{6-8}$$

$$\psi(x,y) = A_x A_y \sin\frac{n_x \pi x}{a}\sin\frac{n_y \pi y}{b} \tag{6-9}$$

$$A_x A_y = \left[\frac{4}{ab}\right]^{1/2} \tag{6-11}$$

$$E_{n_x n_y} = \frac{h^2}{8m}\left[\frac{n_x^2}{a^2} + \frac{n_y^2}{b^2}\right] \tag{6-13}$$

and in analogy with the result of Problem 6-3

$$\text{probability} = \left[\frac{1}{2} + \frac{1}{\pi}\right]^2 = 0.670$$

**6-5.**

$$\text{probability} = \frac{4}{ab}\int_0^{a/2} dx \int_0^{b/2} dy \, \sin^2\left[\frac{n_x \pi x}{a}\right]\sin^2\left[\frac{n_y \pi y}{b}\right]$$

$$= \frac{4}{ab}\left[\frac{a}{4}\right]\left[\frac{b}{4}\right] = \frac{1}{4}$$

The probabilities that the particle is found in either region in Figure 6-17 are the same by symmetry.

**6-6.** The first few energy eigenfunctions of a particle confined to a rectangular region are the same as those sketched in Figure 2-6.

**6-7.**

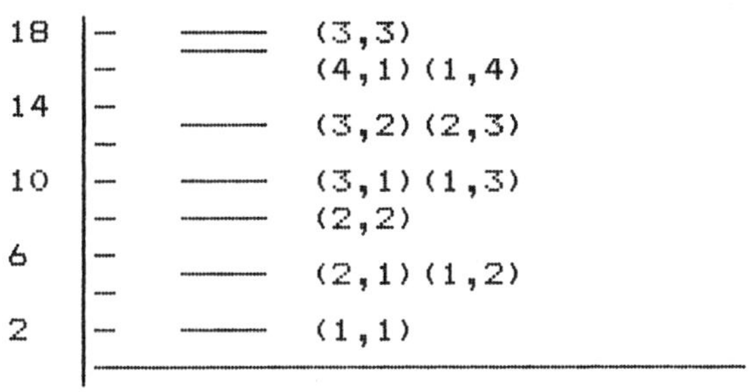


**6-8.** Substitute $\psi(Q_1,Q_2) = \psi_1(Q_1)\psi_2(Q_2)$ into

$$\hat{H}\psi(Q_1,Q_2) = [\hat{H}_1(Q_1) + \hat{H}_2(Q_2)]\psi(Q_1,Q_2)$$
$$= E\psi(Q_1,Q_2)$$

to get

$$\psi_2(Q_2)\hat{H}_1(Q_1)\psi_1(Q_1) + \psi_1(Q_1)\hat{H}(Q_2)\psi_2(Q_2)$$
$$= E_1\psi_1(Q_1)\psi_2(Q_2)) + E_2\psi_1(Q_1)\psi_2(Q_2)$$
$$= (E_1 + E_2)\psi_1(Q_1)\psi_2(Q_2) = (E_1 + E_2)\psi(Q_1,Q_2)$$

**6-9.** The Schrödinger equation in three dimensions for the harmonic oscillator is

$$-\frac{\hbar^2}{2\mu}\left[\frac{\partial^2\psi}{\partial x^2} + \frac{\partial^2\psi}{\partial y^2} + \frac{\partial^2\psi}{\partial z^2}\right] + U(x,y,z)\psi(x,y,z)$$
$$= E\psi(x,y,z)$$

where

$$U(x,y,z) = \frac{1}{2}k_x x^2 + \frac{1}{2}k_y y^2 + \frac{1}{2}k_z z^2$$

Let

$$\psi(x,y,z) = X(x)Y(y)Z(z),$$

$$-\frac{\hbar^2}{2\mu}\left[\frac{\partial^2 X}{\partial x^2}YZ + \frac{\partial^2 Y}{\partial y^2}XZ + \frac{\partial^2 Z}{\partial z^2}XY\right]$$
$$+ U(x,y,z)X(x)Y(y)Z(z)$$

$$= EX(x)Y(y)Z(z)$$

Divide by XYZ and multiply by $(2\mu/\hbar^2)$ to obtain

$$\left[-\frac{1}{X}\frac{\partial^2 X}{\partial x^2} + \frac{2\mu}{\hbar^2}\left[\frac{1}{2}k_x x^2\right]\right] + \left[-\frac{1}{Y}\frac{\partial^2 Y}{\partial y^2} + \frac{2\mu}{\hbar^2}\left[\frac{1}{2}k_y y^2\right]\right]$$
$$+ \left[-\frac{1}{Z}\frac{\partial^2 Z}{\partial z^2} + \frac{2\mu}{\hbar^2}\left[\frac{1}{2}k_z z^2\right]\right]$$
$$= E_x + E_y + E_z = E$$

Now the three-dimensional problem is reduced to three one-dimensional problems, for example

$$\frac{\partial X^2}{\partial x^2} + \frac{2\mu}{\hbar^2}\left[E_x - \frac{1}{2}k_x x^2\right]X(x) = 0$$

The solutions to these harmonic oscillator equations are given in Sections 5-11 through 5-13. Thus,

$$\psi_{n_x n_y n_z}(x,y,z) = \psi_{n_x}(x)\psi_{n_y}(y)\psi_{n_z}(z)$$

where the $\psi_{n_j}$ are harmonic oscillator wave functions and

$$E_{n_x n_y n_z} = h\nu_x\left[n_x + \frac{1}{2}\right] + h\nu_y\left[n_y + \frac{1}{2}\right] + h\nu_z\left[n_z + \frac{1}{2}\right]$$

For an isotropic oscillator

$$E_{n_x n_y n_z} = h\nu(n_x + n_y + n_z + 3/2)$$

and the energy levels are

| Level | States | Degeneracy |
|---|---|---|
| 7 | | |
| 6 | (222) (132) (123) (213) (321) (312) (231) (411) (141) (114) | 10 |
| 5 | (221) (212) (122) (311) (131) (113) | 6 |
| 4 | (211) (121) (112) | 3 |
| 3 | (111) | 1 |

**6-10.** Because

$$r = (x^2 + y^2)^{1/2}$$

$$\left[\frac{\partial r}{\partial x}\right]_y = \frac{2x}{2(x^2 + y^2)^{1/2}} = \frac{x}{r} = \cos\theta$$

$$\left[\frac{\partial r}{\partial y}\right]_x = \frac{2y}{2(x^2 + y^2)^{1/2}} = \frac{y}{r} = \sin\theta$$

Similarly, because

$$\theta = \tan^{-1}\left[\frac{y}{x}\right]$$

we have

$$\left[\frac{\partial \theta}{\partial x}\right]_y = \frac{1}{1 + (y^2/x^2)}\left[-\frac{y}{x^2}\right] = -\frac{y}{x^2 + y^2} = -\frac{\sin\theta}{r}$$

$$\left[\frac{\partial \theta}{\partial y}\right]_x = \frac{1}{1 + (y^2/x^2)}\left[\frac{1}{x}\right] = \frac{x}{x^2 + y^2} = \frac{\cos\theta}{r}$$

Using these in the chain rule expressions given in the problem, we get

$$\left[\frac{\partial f}{\partial x}\right]_y = -\frac{\sin\theta}{r}\left[\frac{\partial f}{\partial \theta}\right]_r$$

$$\left[\frac{\partial f}{\partial y}\right]_x = \frac{\cos\theta}{r}\left[\frac{\partial f}{\partial \theta}\right]_r$$

To solve

$$\frac{d^2\psi}{d\theta^2} + \frac{2\mu r^2 E}{\hbar^2}\psi(\theta) = 0$$

let $\psi(\theta) = e^{\alpha\theta}$ and get

$$\alpha^2 + \frac{2\mu r^2 E}{\hbar^2} = 0$$

or

$$\alpha = \pm i\left[\frac{2\mu r^2 E}{\hbar^2}\right]^{1/2} = \pm ik$$

The solution $\psi(\theta)$ is of the form

$$\psi(\theta) = Ae^{\pm ik\theta}$$

We now require that $\psi(\theta)$ be continuous in the sense that

$$\psi(\theta) = \psi(\theta + 2\pi)$$

This condition gives us that

$$Ae^{\pm ik\theta} = Ae^{\pm ik(\theta+2\pi)} = Ae^{\pm ik\theta}e^{\pm i2\pi k}$$

or that

$$e^{\pm i2\pi k} = 1$$

Use the Euler relation

$$\cos 2\pi k \pm i \sin 2\pi k = 1$$

The solutions to this equation are

$$k = m \qquad\qquad m = 0, \pm 1, \pm 2, \ldots$$

and using the definition of k, we find that

$$E = \frac{m^2\hbar^2}{2\mu r^2} = \frac{m^2\hbar^2}{2I} \qquad\qquad m = 0, \pm 1, \pm 2, \ldots$$

To normalize $\psi(\theta)$, we use

$$A^2\int_0^{2\pi} d\theta\psi^*(\theta)\psi(\theta) = 1$$

to get

$$A = \frac{1}{\sqrt{2\pi}}$$

If we apply this simple model to benzene, we place two of the six $\pi$ electrons in the $m = 0$ level, two in the $m = 1$ level and two in the $m = -1$ level (the $m = 1$ and $m = -1$ levels are degenerate). The energy of the $\pi$ electrons is

$$E_\pi = \frac{4\hbar^2}{2\mu r^2}$$

**6-11.** In this problem we convert

$$\nabla^2 = \frac{\partial^2}{\partial x^2} + \frac{\partial^2}{\partial y^2}$$

to plane polar coordinates. It is a generalization of Problem 6-10 to the case in which the radial coordinate r is not a constant.

Substitute Eqs. 5 into Eqs. 3 and 4 of Problem 6-10.

$$\left[\frac{\partial f}{\partial x}\right]_y = \cos\theta\left[\frac{\partial f}{\partial r}\right]_\theta - \frac{\sin\theta}{r}\left[\frac{\partial f}{\partial \theta}\right]_r$$

$$\left[\frac{\partial f}{\partial y}\right]_x = \sin\theta\left[\frac{\partial f}{\partial r}\right]_\theta + \frac{\cos\theta}{r}\left[\frac{\partial f}{\partial \theta}\right]_r$$

We now apply Eqs. 3 and 4 again

$$\frac{\partial^2 f}{\partial x^2} = \frac{\partial}{\partial x}\left[\frac{\partial f}{\partial x}\right]_y = \left[\frac{\partial r}{\partial x}\right]_y \left[\frac{\partial}{\partial r}\left[\frac{\partial f}{\partial x}\right]_y\right] + \left[\frac{\partial \theta}{\partial x}\right]_y \left[\frac{\partial}{\partial \theta}\left[\frac{\partial f}{\partial x}\right]_y\right]$$

$$= \cos\theta\left[\cos\theta\left[\frac{\partial^2 f}{\partial r^2}\right]_\theta + \frac{\sin\theta}{r^2}\left[\frac{\partial f}{\partial \theta}\right]_r - \frac{\sin\theta}{r}\left[\frac{\partial^2 f}{\partial r\,\partial \theta}\right]\right]$$

$$- \frac{\sin\theta}{r}\left[-\sin\theta\left[\frac{\partial f}{\partial r}\right]_\theta + \cos\theta\left[\frac{\partial^2 f}{\partial \theta\,\partial r}\right]\right.$$

$$\left. - \frac{\cos\theta}{r}\left[\frac{\partial f}{\partial \theta}\right]_r - \frac{\sin\theta}{r}\left[\frac{\partial^2 f}{\partial \theta^2}\right]\right]_r$$

$$= \cos^2\theta\left[\frac{\partial^2 f}{\partial r^2}\right]_\theta - \frac{2\cos\theta\sin\theta}{r}\left[\frac{\partial^2 f}{\partial r\,\partial \theta}\right]$$

$$+ \frac{2\cos\theta\sin\theta}{r}\left[\frac{\partial f}{\partial \theta}\right]_r$$

$$+ \frac{\sin^2\theta}{r}\left[\frac{\partial f}{\partial r}\right]_\theta + \frac{\sin^2\theta}{r^2}\left[\frac{\partial^2 f}{\partial \theta^2}\right]_r$$

and

$$\frac{\partial^2 f}{\partial y^2} = \frac{\partial}{\partial y}\left[\frac{\partial f}{\partial y}\right]_x = \left[\frac{\partial r}{\partial y}\right]_x \left[\frac{\partial}{\partial r}\left[\frac{\partial f}{\partial y}\right]_x\right] + \left[\frac{\partial \theta}{\partial y}\right]_x \left[\frac{\partial}{\partial \theta}\left[\frac{\partial f}{\partial y}\right]_x\right]$$

$$= \sin\theta\left[\sin\theta\left[\frac{\partial^2 f}{\partial r^2}\right]_\theta - \frac{\cos\theta}{r^2}\left[\frac{\partial f}{\partial \theta}\right]_r + \frac{\cos\theta}{r}\left[\frac{\partial^2 f}{\partial r\,\partial \theta}\right]\right]$$

$$+ \frac{\cos\theta}{r}\left[\cos\theta\left[\frac{\partial f}{\partial r}\right]_\theta + \sin\theta\left[\frac{\partial^2 f}{\partial \theta\,\partial r}\right]\right.$$

$$\left. - \frac{\sin\theta}{r}\left[\frac{\partial f}{\partial \theta}\right]_r + \frac{\cos\theta}{r}\left[\frac{\partial^2 f}{\partial \theta^2}\right]\right]_r$$

$$= \sin^2\theta\left[\frac{\partial^2 f}{\partial r^2}\right]_\theta + \frac{2\cos\theta\sin\theta}{r}\left[\frac{\partial^2 f}{\partial r\,\partial \theta}\right]$$

$$- \frac{2\cos\theta\sin\theta}{r}\left[\frac{\partial f}{\partial \theta}\right]_r$$

$$+ \frac{\cos^2\theta}{r}\left[\frac{\partial f}{\partial r}\right]_\theta + \frac{\cos^2\theta}{r^2}\left[\frac{\partial^2 f}{\partial \theta^2}\right]_r$$

$$\frac{\partial^2 f}{\partial x^2} + \frac{\partial^2 f}{\partial y^2} = \frac{\partial^2 f}{\partial r^2} + \frac{1}{r}\frac{\partial f}{\partial r} + \frac{1}{r^2}\frac{\partial^2 f}{\partial \theta^2}$$

Consider the Schrödinger equation of a particle moving under a central potential in two dimensions

$$-\frac{\hbar^2}{2\mu}\left[\frac{\partial^2 \psi}{\partial r^2} + \frac{1}{r}\frac{\partial \psi}{\partial r} + \frac{1}{r^2}\frac{\partial^2 \psi}{\partial \theta^2}\right] + U(r)\psi(r,\theta) = E\psi(r,\theta)$$

Multiply through by $2\mu r^2$ to get

$$-\hbar^2\left[r^2\frac{\partial^2\psi}{\partial r^2} + r\frac{\partial\psi}{\partial r}\right] - \hbar^2\frac{\partial^2\psi}{\partial\theta^2} + 2\mu r^2[U(r) - E]\psi = 0$$

Now let

$$\psi(r,\theta) = R(r)\Theta(\theta)$$

to get

$$-\frac{\hbar^2}{R(r)}\left[r^2\frac{d^2R}{dr^2} + r\frac{dR}{dr}\right] + 2\mu r^2[U(r)-E] - \frac{\hbar^2}{\Theta}\frac{d^2\Theta}{d\theta^2} = 0$$

and assume that (Problem 6-10)

$$\Theta(\theta) = (2\pi)^{-1/2}\, e^{im\theta}$$

to get

$$-\frac{\hbar^2}{R(r)}\left[r^2\frac{d^2R}{dr^2} + r\frac{dR}{dr}\right] + 2\mu r^2[U(r) - E] + m^2\hbar^2$$

or

$$-\frac{\hbar^2}{2\mu}\left[\frac{d^2R}{dr^2} + \frac{1}{r}\frac{dR}{dr}\right] + \left[\frac{m^2\hbar^2}{2\mu r^2} + U(r) - E\right]R(r) = 0$$

This result is the two dimensional analog of Eq. 6-98.

**6-12.** Eq. 6-64 gives

$$\nu = \frac{h}{4\pi^2 I}(\ell + 1) \qquad\qquad \ell = 0,1,2,\ldots$$

for the frequencies associated with rotational transitions. To estimate typical values of $\nu$, we must first estimate I, the moment of inertia of a diatomic molecule. Using typical values of $\mu$ and r,

$$\begin{aligned} I &= \mu r^2 \\ &= (10\ \text{amu})(1.66\times10^{-27}\ \text{kg}\cdot\text{amu}^{-1})(3\times10^{-10}\ \text{m})^2 \\ &= 1.5\times10^{-45}\ \text{kg}\cdot\text{m}^2 \end{aligned}$$

Thus, the observed frequency is an integral multiple of

$$\nu = \frac{6.626\times10^{-34}\ \text{J}\cdot\text{s}}{4\pi^2(1.5\times10^{-45}\ \text{kg}\cdot\text{m}^2)} = 1.1\times10^{10}\ \text{s}^{-1}$$

Referring to Figure 1-14, we see that these frequencies occur in the microwave or far infrared region.

**6-13.** This problem is similar to Example 6-5. According to Eq. 6-67,

$$\bar{\nu} = 2\bar{B}(\ell + 1) \qquad \ell = 0,1,2,\ldots$$

where

$$\bar{B} = \frac{h}{8\pi^2 cI}$$

Thus, we have

$$16.72\ \text{cm}^{-1} = 2\bar{B} = \frac{2h}{8\pi^2 cI}$$

Solving for I

$$\begin{aligned} I &= \frac{2h}{8\pi^2 c(2\bar{B})} \\ &= \frac{2(6.626 \times 10^{-34}\ \text{J·s})}{8\pi^2(2.998 \times 10^{10}\ \text{cm·s}^{-1})(16.72\ \text{cm}^{-1})} \\ &= 3.35 \times 10^{-47}\ \text{kg·m}^2 \end{aligned}$$

Solving $I = \mu r^2$ for r, we obtain

$$\begin{aligned} r &= \left[\frac{I}{\mu}\right]^{1/2} \\ &= \left[\frac{3.35 \times 10^{-47}\ \text{kg·m}^2}{((1\times79)/80\ \text{amu})(1.6607\times10^{-27}\ \text{kg·amu}^{-1})}\right]^{1/2} \\ &= 1.43 \times 10^{-10}\ \text{m} = 143\ \text{pm} \end{aligned}$$

**6-14.** This problem is similar to Example 6-5 and Problem 6-13. We use Eq. 6-65

$$\nu = 2B(\ell + 1) \qquad \ell = 0,1,2,\ldots$$

where

$$B = \frac{h}{8\pi^2 I}$$

$$\nu = 2B = 1.153 \times 10^{11}\ \text{s}^{-1} = \frac{h}{4\pi^2 I}$$

Solving for I

$$I = \frac{6.626 \times 10^{-34}\ \text{J·s}}{4\pi^2(1.153 \times 10^{11}\ \text{s}^{-1})} = 1.456 \times 10^{-46}\ \text{kg·m}^2$$

Using $I = \mu r^2$, we find

$$r = \left[\frac{I}{\mu}\right]^{1/2}$$

$$= \left[\frac{1.456 \times 10^{-46}\ \text{kg}\cdot\text{m}^2}{((12\times 16)/28\ \text{amu})(1.6607\times 10^{-27}\ \text{kg}\cdot\text{amu}^{-1})}\right]^{1/2}$$

$$= 1.13 \times 10^{-10}\ \text{m} = 113\ \text{pm}$$

**6-15.** Legendre's equation is

$$\sin\theta\frac{d}{d\theta}\left[\sin\theta\frac{d\Theta}{d\theta}\right] + (\beta\sin^2\theta - m^2)\Theta(\theta) = 0$$

or

$$\sin^2\theta\frac{d^2\Theta}{d\theta^2} + \sin\theta\cos\theta\frac{d\Theta}{d\theta} + (\beta\sin^2\theta - m^2)\Theta(\theta) = 0$$

Let $x = \cos\theta$ and $\Theta(\theta) = P(x)$.

$$\frac{d\Theta}{d\theta} = \frac{dP}{dx}\frac{dx}{d\theta} = -\sin\theta\frac{dP}{dx} = -(1 - x^2)^{1/2}\frac{dP}{dx}$$

$$\frac{d^2\Theta}{d\theta^2} = \frac{d}{d\theta}\left[\frac{d\Theta}{d\theta}\right] = \left[\frac{dx}{d\theta}\right]\frac{d}{dx}\left[-(1 - x^2)^{1/2}\frac{dP}{dx}\right]$$

$$= -\sin\theta\left[\frac{x}{(1 - x^2)^{1/2}}\frac{dP}{dx} - (1 - x^2)^{1/2}\frac{d^2P}{dx^2}\right]$$

$$= -x\frac{dP}{dx} + (1 - x^2)\frac{d^2P}{dx^2}$$

Substituting these results into Legendre's equation gives

$$(1 - x^2)^2\frac{d^2P}{dx^2} - 2x(1 - x^2)\frac{dP}{dx} + (\beta(1 - x^2) - m^2)P(x) = 0$$

Divide through by $(1 - x^2)$ and get

$$(1 - x^2)\frac{d^2P}{dx^2} - 2x\frac{dP}{dx} + \left[\beta - \frac{m^2}{1 - x^2}\right]P(x) = 0$$

**6-16.** Eq. 6-69 with $m = 0$ is

$$(1 - x^2)\frac{d^2P}{dx^2} - 2x\frac{dP}{dx} + \ell(\ell + 1)P(x) = 0$$

For $\ell = 0$, $P_0(x) = 1$ and the equation is clearly satisfied.

For $\ell = 1$, $P_1(x) = x$, and we have

$$0 - 2x(1) + (1\cdot 2)(x) = 0$$

For $\ell = 2$, $P_2(x) = (1/2)(3x^2 - 1)$, and

$$(1 - x^2)(3) - 2x(3x) + (2\cdot 3)[(1/2)(3x^2 - 1)] = 0$$

For $\ell = 3$, $P_3(x) = (1/2)(5x^3 - 3x)$, and

$$(1 - x^2)(15x) - 2x[(1/2)(15x^2 - 3)] + (3\cdot 4)[(1/2)(5x^3 - 3x)] = 0$$

For $\ell = 4$, $P_4(x) = (1/8)(35x^4 - 30x^2 + 3)$, and

$$(1 - x^2)[(1/2)(105x^2 - 15)] - 2x[(1/2)(35x^3 - 15x)] + (4\cdot 5)[(1/8)(35x^4 - 30x^2 + 3)] = 0$$

**6-17.** Eq. 6-70 is

$$\int_{-1}^{1} dx P_\ell(x) P_n(x) = 0 \qquad \ell \neq n$$

Let $x = \cos\theta$, so that $dx = -\sin\theta d\theta$ and the limits become $\pi$ to 0

$$-\int_{\pi}^{0} P_\ell(\cos\theta) P_n(\cos\theta) \sin\theta d\theta = \int_{0}^{\pi} P_\ell(\cos\theta) P_n(\cos\theta) \sin\theta d\theta = 0 \qquad \ell \neq n$$

**6-18.** Eqs. 6-70 and 6-71 can be written together as

$$\int_{-1}^{1} P_\ell(x) P_n(x) dx = \frac{2\delta_{\ell n}}{2\ell + 1}$$

Some examples are (for $\ell = 0$, 1 and 2)

$$\int_{-1}^{1} P_0^2(x) dx = \int_{-1}^{1} dx = 2$$

$$\int_{-1}^{1} P_1^2(x) dx = \int_{-1}^{1} x^2 dx = \frac{2}{3}$$

$$\int_{-1}^{1} P_2^2(x)\,dx = \frac{1}{4}\int_{-1}^{1}(3x^2 - 1)^2 dx$$

$$= \frac{1}{4}\left[\frac{9\cdot 2}{5} - \frac{6\cdot 2}{3} + 2\right] = \frac{2}{5}$$

$$\int_{-1}^{1} P_0(x)P_1(x)\,dx = \int_{-1}^{1} x\ dx = 0$$

$$\int_{-1}^{1} P_0(x)P_2(x)\,dx = \frac{1}{2}\int_{-1}^{1}(3x^2 - 1)\,dx = 1 - 1 = 0$$

$$\int_{-1}^{1} P_1(x)P_2(x)\,dx = \frac{1}{2}\int_{-1}^{1} x(3x^2 - 1)\,dx = 0$$

**6-19.** Eq. 6-72 is

$$P_\ell^{|m|}(x) = (1 - x^2)^{|m|/2}\frac{d^m}{dx^m}P_\ell(x)$$

$$P_0^0(x) = (1 - x^2)^0\frac{d^0}{dx^0}P_0(x) = P_0(x) = 1$$

$$P_1^0(x) = (1 - x^2)^0\frac{d^0}{dx^0}P_1(x) = P_1(x) = x$$

$$P_1^1(x) = (1 - x^2)^{1/2}\frac{d}{dx}P_1(x)$$

$$= (1 - x^2)^{1/2}(1) = (1 - x^2)^{1/2}$$

$$P_2^0(x) = (1 - x^2)^0\frac{d^0}{dx^0}P_2(x)$$

$$= P_2(x) = \frac{1}{2}(3x^2 - 1)$$

$$P_2^1(x) = (1 - x^2)^{1/2}\frac{d}{dx}P_2(x)$$

$$= (1 - x^2)^{1/2}\frac{d}{dx}\left[\frac{1}{2}(3x^2 - 1)\right]$$

$$= 3x(1 - x^2)^{1/2}$$

$$P_2^2(x) = (1 - x^2)\frac{d^2}{dx^2}P_2(x)$$

$$= (1 - x^2)\frac{d^2}{dx^2}\left[\frac{1}{2}(3x^2 - 1)\right]$$

$$= 3(1 - x^2)$$

$$P_3^0(x) = (1 - x^2)^0 \frac{d^0}{dx^0} P_3(x)$$

$$= P_3(x) = \frac{1}{2}(5x^3 - 3x)$$

$$P_3^1(x) = (1 - x^2)^{1/2} \frac{d}{dx} P_3(x)$$

$$= (1 - x^2)^{1/2} \frac{d}{dx}\left[\frac{1}{2}(5x^3 - 3x)\right]$$

$$= \frac{3}{2}(5x^2 - 1)(1 - x^2)^{1/2}$$

$$P_3^2(x) = (1 - x^2)\frac{d^2}{dx^2} P_3(x)$$

$$= (1 - x^2)\frac{d^2}{dx^2}\left[\frac{1}{2}(5x^3 - 3x)\right]$$

$$= 15x(1 - x^2)$$

$$P_3^3(x) = (1 - x^2)^{3/2} \frac{d^3}{dx^3} P_3(x)$$

$$= (1 - x^2)^{3/2}\frac{d^3}{dx^3}\left[\frac{1}{2}(5x^3 - 3x)\right]$$

$$= 15(1 - x^2)^{3/2}$$

**6-20.** Eq. 6-69 is

$$(1 - x^2)\frac{d^2P_\ell^{|m|}}{dx^2} - 2x\frac{dP_\ell^{|m|}}{dx}$$

$$+ \left[\ell(\ell + 1) - \frac{m^2}{1 - x^2}\right]P_\ell^{|m|}(x) = 0$$

When $\ell = m = 0$, $P_0^0(x) = 1$ clearly satisfies this equation.

When $\ell = 1$ and $m = 0$, $P_1^0(x) = x$, and we have

$$0 - 2x(1) + (2 - 0)x = 0$$

When $\ell = 1$ and $m = 1$, $P_1^1(x) = (1 - x^2)^{1/2}$, and

$$(1 - x^2)\left[- \frac{x^2}{(1 - x^2)^{3/2}} - \frac{1}{(1 - x^2)^{1/2}}\right]$$

$$- 2x\left[- \frac{x}{(1 - x^2)^{1/2}}\right]$$

$$+ \left[2 - \frac{1}{1 - x^2}\right](1 - x^2)^{1/2}$$

$$= -\frac{x^2}{(1-x^2)^{1/2}} - (1-x^2)^{1/2} + \frac{2x^2}{(1-x^2)^{1/2}}$$

$$+ 2(1-x^2)^{1/2} - \frac{1}{(1-x^2)^{1/2}}$$

$$= \frac{x^2-1}{(1-x^2)^{1/2}} + (1-x^2)^{1/2}$$

$$= 0$$

The orthonormality condition is (Eq. 6-74)

$$\int_{-1}^{1} dx P_\ell^{|m|}(x) P_\ell^{|m|}(x) = \frac{2}{(2\ell+1)} \frac{(\ell+|m|)!}{(\ell-|m|)!} \delta_{\ell n}$$

For example

$$\int_{-1}^{1} dx [P_1^1(x)]^2 = \int_{-1}^{1} dx (1-x^2) = \frac{4}{3} = \frac{2\ (2!)}{(3)\ (0!)}$$

$$\int_{-1}^{1} dx [P_2^1(x)]^2 = \int_{-1}^{1} dx (9x^2 - 9x^4) = \frac{12}{5} = \frac{2}{(5)} \frac{(3!)}{(1!)}$$

and

$$\int_{-1}^{1} dx P_1^1(x) P_2^1(x) = \int_{-1}^{1} dx [3x(1-x^2)] = 0$$

**6-21.** The recursion formula is

$$(2\ell+1)xP_\ell^{|m|}(x)$$

$$= (\ell - |m| + 1)P_{\ell+1}^{|m|}(x) + (\ell + |m|)P_{\ell-1}^{|m|}(x)$$

Let $\ell = 1$ and $m = 0$

$$3xP_1^0(x) \stackrel{?}{=} 2P_2^0(x) + P_0^0(x)$$

$$3x^2 = (3x^2 - 1) + 1$$

Let $\ell = 1$ and $m = 1$

$$3xP_1^1(x) \stackrel{?}{=} P_2^1(x) + 2P_0^1(x)$$

$$3x(1-x^2)^{1/2} = 3x(1-x^2)^{1/2} = 0$$

where we have set $P_0^1(x)$ equal to zero because m cannot exceed $\ell$.

Let $\ell = 2$ and $m = 0$

$$5xP_2^0(x) \stackrel{?}{=} 3P_3^0(x) + 2P_1^0(x)$$

$$(5/2)(3x^3 - x) = (3/2)(5x^3 - 3x) + 2x$$

Let $\ell = 2$ and $m = 1$

$$5xP_2^1(x) \stackrel{?}{=} 2P_3^1(x) + 3P_1^1(x)$$

$$15x^2(1 - x^2)^{1/2}$$
$$= (15x^2 - 3)(1 - x^2)^{1/2} + 3(1 - x^2)^{1/2}$$

Let $\ell = 2$ and $m = 2$

$$5xP_2^2(x) \stackrel{?}{=} P_3^2(x) + 4P_1^2(x)$$

$$15x(1 - x^2) = 15x(1 - x^2) + 0$$

where we have set $P_1^2(x)$ equal to zero because $m$ cannot exceed $\ell$.

**6-22.** Eq. 6-77 is

$$\int_0^{2\pi} d\phi \int_0^{\pi} d\theta \sin\theta \; Y_\ell^m(\theta,\phi)^* Y_n^k(\theta,\phi) = \delta_{n\ell}\delta_{mk}$$

For example, the normalization condition is

$$\int_0^{2\pi} d\phi \int_0^{\pi} d\theta \sin\theta |Y_0^0(\theta,\phi|^2 = \int_0^{2\pi} d\phi \int_0^{\pi} d\theta \sin\theta \left[\frac{1}{4\pi}\right] = 1$$

$$\int_0^{2\pi} d\phi \int_0^{\pi} d\theta \sin\theta |Y_1^0(\theta,\phi|^2 = \int_0^{2\pi} d\phi \int_0^{\pi} d\theta \sin\theta \left[\frac{3\cos^2\theta}{4\pi}\right]$$

$$= \frac{3}{2}\int_{-1}^{1} dx \; x^2 = 1$$

$$\int_0^{2\pi} d\phi \int_0^{\pi} d\theta \sin\theta |Y_1^1(\theta,\phi|^2 = \int_0^{2\pi} d\phi \int_0^{\pi} d\theta \sin\theta |Y_1^{-1}(\theta,\phi)|^2$$

$$= \int_0^{2\pi} d\phi \int_0^{\pi} d\theta \sin\theta \left[\frac{3\sin^2\theta}{8\pi}\right]$$

$$= \frac{3}{4}\int_{-1}^{1} dx(1 - x^2) = 1$$

and the orthogonality condition is

$$\int_0^{2\pi} d\phi \int_0^{\pi} d\theta \sin\theta \; Y_0^0(\theta,\phi)^* Y_1^0(\theta,\phi)$$

$$= \int_0^{2\pi} d\phi \int_0^{\pi} d\theta \sin\theta \left[\frac{3^{1/2}\cos\theta}{4\pi}\right]$$

$$= \frac{3^{1/2}}{2}\int_{-1}^{1} dx \; x = 0$$

$$\int_0^{2\pi} d\phi \int_0^{\pi} d\theta \sin\theta \; Y_0^0(\theta,\phi)^* Y_1^{\pm 1}(\theta,\phi)$$

$$= \int_0^{2\pi} d\phi \int_0^{\pi} d\theta \sin\theta \left[\frac{3^{1/2} e^{\pm i\phi}\cos\theta}{2^{1/2}\, 4\pi}\right]$$

$= 0$ because of the integration over $\phi$

$$\int_0^{2\pi} d\phi \int_0^{\pi} d\theta \sin\theta \; Y_1^0(\theta,\phi)^* Y_1^{\pm 1}(\theta,\phi)$$

$$= \int_0^{2\pi} d\phi \int_0^{\pi} d\theta \sin\theta \left[\frac{3^{1/2} e^{\pm i\phi}\sin\theta\cos\theta}{2^{1/2}\, 4\pi}\right]$$

$= 0$ because of the integration over $\phi$

**6-23.**

$$|Y_1^1(\theta,\phi)|^2 + |Y_1^0(\theta,\phi)|^2 + |Y_1^{-1}(\theta,\phi)|^2$$

$$= \frac{3\sin^2\theta}{8\pi} + \frac{3\cos^2\theta}{4\pi} + \frac{3\sin^2\theta}{8\pi} = \frac{3}{4\pi}$$

**6-24.** We wish to convert

$$\hat{L}_z = -i\hbar\left[x\frac{\partial}{\partial y} - y\frac{\partial}{\partial x}\right]$$

to spherical coordinates. Spherical coordinates are given by

$$x = r\sin\theta\cos\phi$$

$$y = r\sin\theta\cos\phi$$

$$z = r\cos\theta$$

or

$$r = (x^2 + y^2 + z^2)^{1/2}$$

$$\phi = \tan^{-1}(y/x)$$

$$\theta = \cos^{-1}\left[\frac{z}{(x^2 + y^2 + z^2)^{1/2}}\right]$$

To carry out the conversion, we use the chain rule of partial differentiation (see also Problems 6-10 and 6-11).

$$\left[\frac{\partial f}{\partial x}\right] = \left[\frac{\partial f}{\partial r}\right]\left[\frac{\partial r}{\partial x}\right] + \left[\frac{\partial f}{\partial \theta}\right]\left[\frac{\partial \theta}{\partial x}\right] + \left[\frac{\partial f}{\partial \phi}\right]\left[\frac{\partial \phi}{\partial x}\right]$$

$$\left[\frac{\partial f}{\partial y}\right] = \left[\frac{\partial f}{\partial r}\right]\left[\frac{\partial r}{\partial y}\right] + \left[\frac{\partial f}{\partial \theta}\right]\left[\frac{\partial \theta}{\partial y}\right] + \left[\frac{\partial f}{\partial \phi}\right]\left[\frac{\partial \phi}{\partial y}\right]$$

Using the above relations between spherical coordinates and cartesian coordinates, we get

$$\left[\frac{\partial f}{\partial x}\right] = \sin\theta\cos\phi\left[\frac{\partial f}{\partial r}\right] + \frac{\cos\theta\cos\phi}{r}\left[\frac{\partial f}{\partial \theta}\right] - \frac{\sin\phi}{r\sin\theta}\left[\frac{\partial f}{\partial \phi}\right]$$

$$\left[\frac{\partial f}{\partial y}\right] = \sin\theta\sin\phi\left[\frac{\partial f}{\partial r}\right] + \frac{\cos\theta\sin\phi}{r}\left[\frac{\partial f}{\partial \theta}\right] + \frac{\cos\phi}{r\sin\theta}\left[\frac{\partial f}{\partial \phi}\right]$$

Now,

$$x\left[\frac{\partial f}{\partial y}\right] - y\left[\frac{\partial f}{\partial x}\right]$$

$$= (r\sin^2\theta\cos\phi\sin\phi - r\sin^2\theta\cos\phi\sin\phi)\left[\frac{\partial f}{\partial r}\right]$$

$$+ (\sin\theta\cos\theta\cos\phi\sin\phi - \sin\theta\cos\theta\sin\phi\cos\phi)\left[\frac{\partial f}{\partial \theta}\right]$$

$$+ (\cos^2\phi + \sin^2\phi)\left[\frac{\partial f}{\partial \phi}\right]$$

$$= \frac{\partial f}{\partial \phi}$$

Therefore,

$$\hat{L}_z = -i\hbar\frac{\partial}{\partial \phi}$$

**6-25.** Much of this problem is done in Problem 6-24 and many of the necessary quantities are given there. The only new quantity that we need is

$$\left[\frac{\partial f}{\partial z}\right] = \left[\frac{\partial f}{\partial r}\right]\left[\frac{\partial r}{\partial z}\right] + \left[\frac{\partial f}{\partial \theta}\right]\left[\frac{\partial \theta}{\partial z}\right] + \left[\frac{\partial f}{\partial \phi}\right]\left[\frac{\partial \phi}{\partial z}\right]$$

$$= \cos\theta\left[\frac{\partial f}{\partial r}\right] - \frac{\sin\theta}{r}\left[\frac{\partial f}{\partial \theta}\right] + 0\left[\frac{\partial f}{\partial \phi}\right]$$

Therefore,

$$\hat{L}_x = -i\hbar\left[y\frac{\partial}{\partial z} - z\frac{\partial}{\partial y}\right]$$

$$= -i\hbar\left[(r\sin\theta\cos\theta\sin\phi - r\sin\theta\cos\theta\sin\phi)\frac{\partial}{\partial r}\right.$$
$$+ (-\sin^2\theta\sin\phi - \cos^2\theta\sin\phi)\frac{\partial}{\partial \theta}$$
$$\left. + (0 - \cot\theta\cos\phi)\frac{\partial}{\partial \phi}\right]$$

$$= -i\hbar\left[-\sin\phi\frac{\partial}{\partial \theta} - \cot\theta\cos\phi\frac{\partial}{\partial \phi}\right]$$

and

$$\hat{L}_y = -i\hbar\left[z\frac{\partial}{\partial x} - x\frac{\partial}{\partial z}\right]$$

$$= -i\hbar\left[(r\cos\theta\sin\theta\cos\phi - r\cos\theta\sin\theta\cos\phi)\frac{\partial}{\partial r}\right.$$
$$+ (\cos^2\theta\cos\phi + \sin^2\theta\cos\phi)\frac{\partial}{\partial \theta}$$
$$\left. + (-\cot\theta\sin\phi - 0)\frac{\partial}{\partial \phi}\right]$$

$$= -i\hbar\left[\cos\phi\frac{\partial}{\partial \theta} - \cot\theta\sin\phi\frac{\partial}{\partial \phi}\right]$$

**6-26.**

$$[\hat{L}_x,\hat{L}_y]f = (\hat{L}_x\hat{L}_y - \hat{L}_y\hat{L}_x)f$$

$$= -\hbar^2\left[\left[y\frac{\partial}{\partial z} - z\frac{\partial}{\partial y}\right]\left[z\frac{\partial f}{\partial x} - x\frac{\partial f}{\partial z}\right]\right.$$
$$\left. - \left[z\frac{\partial}{\partial x} - x\frac{\partial}{\partial z}\right]\left[y\frac{\partial f}{\partial z} - z\frac{\partial f}{\partial y}\right]\right]$$

$$= -\hbar^2\left[y\frac{\partial f}{\partial x} + yz\frac{\partial^2 f}{\partial x\,\partial z} - xy\frac{\partial^2 f}{\partial z^2}\right.$$
$$- z^2\frac{\partial^2 f}{\partial x\,\partial y} + xz\frac{\partial^2 f}{\partial x\,\partial z} - yz\frac{\partial^2 f}{\partial x\,\partial z} + z^2\frac{\partial^2 f}{\partial x\,\partial y}$$
$$\left. + xy\frac{\partial^2 f}{\partial z^2} - x\frac{\partial f}{\partial y} - xz\frac{\partial^2 f}{\partial y\,\partial z}\right]$$

$$= -\hbar^2\left[y\frac{\partial f}{\partial x} - x\frac{\partial f}{\partial y}\right]$$

and so

$$[\hat{L}_y,\hat{L}_z] = i\hbar\left[-i\hbar\left[x\frac{\partial}{\partial y} - y\frac{\partial}{\partial x}\right]\right] = i\hbar\hat{L}_z$$

Similarly,

$$[\hat{L}_y,\hat{L}_z]f = (\hat{L}_y\hat{L}_z - \hat{L}_z\hat{L}_y)f$$

$$= -\hbar^2\left[\left[z\frac{\partial}{\partial x} - x\frac{\partial}{\partial z}\right]\left[x\frac{\partial f}{\partial y} - y\frac{\partial f}{\partial x}\right] - \left[x\frac{\partial}{\partial y} - y\frac{\partial}{\partial x}\right]\left[z\frac{\partial f}{\partial x} - x\frac{\partial f}{\partial z}\right]\right]$$

$$= -\hbar^2\left[z\frac{\partial f}{\partial y} + xz\frac{\partial^2 f}{\partial x\,\partial y} - yz\frac{\partial^2 f}{\partial x^2} - x^2\frac{\partial^2 f}{\partial y\,\partial z} + xy\frac{\partial^2 f}{\partial x\,\partial z} - xz\frac{\partial^2 f}{\partial x\,\partial y} + x^2\frac{\partial^2 f}{\partial y\,\partial z} + yz\frac{\partial^2 f}{\partial x^2} - y\frac{\partial f}{\partial z} - xy\frac{\partial^2 f}{\partial x\,\partial z}\right]$$

$$= -\hbar^2\left[z\frac{\partial f}{\partial y} - y\frac{\partial f}{\partial z}\right]$$

and so

$$[\hat{L}_y,\hat{L}_z] = i\hbar\left[-i\hbar\left[y\frac{\partial}{\partial z} - z\frac{\partial}{\partial y}\right]\right] = i\hbar\hat{L}_x$$

Lastly,

$$[\hat{L}_z,\hat{L}_x]f = (\hat{L}_z\hat{L}_x - \hat{L}_x\hat{L}_z)f$$

$$= -\hbar^2\left[\left[x\frac{\partial}{\partial y} - y\frac{\partial}{\partial x}\right]\left[y\frac{\partial f}{\partial z} - z\frac{\partial f}{\partial y}\right] - \left[y\frac{\partial}{\partial z} - z\frac{\partial}{\partial y}\right]\left[x\frac{\partial f}{\partial y} - y\frac{\partial f}{\partial x}\right]\right]$$

$$= -\hbar^2\left[x\frac{\partial f}{\partial z} + xy\frac{\partial^2 f}{\partial y\,\partial z} - xz\frac{\partial^2 f}{\partial y^2} - y^2\frac{\partial^2 f}{\partial x\,\partial z} + yz\frac{\partial^2 f}{\partial x\,\partial y} - xy\frac{\partial^2 f}{\partial y\,\partial z} + y^2\frac{\partial^2 f}{\partial x\,\partial z} + xz\frac{\partial^2 f}{\partial y^2} - z\frac{\partial f}{\partial x} - yz\frac{\partial^2 f}{\partial x\,\partial y}\right]$$

$$= -\hbar^2\left[x\frac{\partial f}{\partial z} - z\frac{\partial f}{\partial x}\right]$$

and so

$$[\hat{L}_z,\hat{L}_x] = i\hbar\left[-i\hbar\left[z\frac{\partial}{\partial x} - x\frac{\partial}{\partial z}\right]\right] = i\hbar\hat{L}_y$$

The proof that $\hat{L}^2$ commutes with $\hat{L}_z$ is given in Example 6-8. Since the labelling of x, y, and z is arbitrary, $\hat{L}^2$ must also commute with $\hat{L}_x$ and $\hat{L}_y$.

**6-27.**

$$\langle L_x \rangle = \int_0^{2\pi} d\phi \int_0^{\pi} d\theta \sin\theta \; Y_\ell^m(\theta,\phi)^* \hat{L}_x Y_\ell^m(\theta,\phi)$$

where (Problem 6-25)

$$\hat{L}_x = -i\hbar\left[-\sin\phi\frac{\partial}{\partial\theta} - \cot\theta\cos\phi\frac{\partial}{\partial\phi}\right]$$

$\langle L_x \rangle$ is zero for $\ell = m = 0$ because $Y_0^0(\theta,\phi)$ = constant.

For $\ell = 1$ and $m = 0$,

$$\hat{L}_x Y_1^0(\theta,\phi)$$

$$= -i\hbar\left[-\sin\phi\frac{\partial}{\partial\theta} - \cot\theta\cos\phi\frac{\partial}{\partial\phi}\right]\left[\left[\frac{3}{4\pi}\right]^{1/2}\cos\theta\right]$$

$$= -i\hbar\left[\frac{3}{4\pi}\right]^{1/2}\sin\phi\sin\theta$$

and

$$\langle L_x \rangle = -i\hbar\left[\frac{3}{4\pi}\right]\int_0^{2\pi} d\phi\sin\phi \int_0^{\pi} d\theta\sin^2\theta\cos\theta$$

$$= 0 \text{ due to the integration over } \phi$$

For $\ell = 1$ and $m = \pm 1$

$$\hat{L}_x Y_1^{\pm 1}(\theta,\phi)$$

$$= -i\hbar\left[-\sin\phi\frac{\partial}{\partial\theta} - \cot\theta\cos\phi\frac{\partial}{\partial\phi}\right]\left[\left[\frac{3}{8\pi}\right]^{1/2} e^{\pm i\phi}\sin\theta\right]$$

$$= i\hbar\left[\frac{3}{8\pi}\right]^{1/2}[e^{\pm i\phi}\sin\phi\cos\theta \pm ie^{\pm i\phi}\cos\theta\cos\phi]$$

and

$$\langle L_x \rangle = i\hbar\left[\frac{3}{8\pi}\right]\int_0^{2\pi} d\phi \int_0^{\pi} d\theta\sin\theta[\sin\phi\cos\theta \pm i\cos\phi\cos\theta]$$

$$= 0$$

because

$$\int_0^{2\pi} d\phi\sin\phi = \int_0^{2\pi} d\phi\cos\phi = 0$$

Since the labelling of x and y is arbitrary, we must have that $\langle L_y \rangle = 0$ also.

**6-28.** The orthonormality condition (Eq. 6-105) is

$$\int_0^{2\pi} d\phi \int_0^{\pi} d\theta \sin\theta \int_0^{\infty} dr\ r^2 \psi^*_{n\ell m}(r,\theta,\phi)\psi_{n'\ell'm'}(r,\theta,\phi)$$

$$= \delta_{nn'}\delta_{\ell\ell'}\delta_{mm'}$$

We first show that the first few hydrogen atomic wave functions in Table 6-5 are normalized.

$$\int_0^{2\pi} d\phi \int_0^{\pi} d\theta \sin\theta \int_0^{\infty} dr\ r^2 \left[\frac{1}{\pi}\left[\frac{Z}{a_0}\right]^3 e^{-2\sigma}\right]$$

$$= 4\left[\frac{Z}{a_0}\right]^3 \int_0^{\infty} \exp(-2Zr/a_0) r^2 dr = 4\int_0^{\infty} e^{-2\sigma}\sigma^2 d\sigma = 1$$

$$\int_0^{2\pi} d\phi \int_0^{\pi} d\theta \sin\theta \int_0^{\infty} dr\ r^2 \left[\frac{1}{32\pi}\left[\frac{Z}{a_0}\right]^3 (4 - 4\sigma + \sigma^2) e^{-\sigma}\right]$$

$$= \frac{1}{8}\left[\frac{Z}{a_0}\right]^3 \int_0^{\infty} dr\ r^2 (4 - 4\sigma + \sigma^2) e^{-\sigma}$$

$$= \frac{1}{8}\int_0^{\infty} d\sigma (4\sigma^2 - 4\sigma^3 + \sigma^4) e^{-\sigma}$$

$$= \frac{1}{8}[4 \cdot 2! - 4 \cdot 3! + 4!] = 1$$

$$\int_0^{2\pi} d\phi \int_0^{\pi} d\theta \sin\theta \int_0^{\infty} dr\ r^2 \left[\frac{1}{32\pi}\left[\frac{Z}{a_0}\right]^3 \sigma^2 e^{-\sigma} \cos^2\theta\right]$$

$$= \frac{1}{16}\left[\frac{Z}{a_0}\right]^3 \int_0^{\pi} d\theta \sin\theta \cos^2\theta \int_0^{\infty} dr\ r^2 \sigma^2 e^{-\sigma}$$

$$= \frac{1}{16} \int_{-1}^{1} dx\ x^2 \int_0^{\infty} d\sigma\ \sigma^4 e^{-\sigma} = \frac{1}{16}\left[\frac{2}{3}\right]\left[4!\right] = 1$$

and finally

$$\int_0^{2\pi} d\phi \int_0^{\pi} d\theta \sin\theta \int_0^{\infty} dr\ r^2 \left[\frac{1}{64\pi}\left[\frac{Z}{a_0}\right]^3 \sigma^2 e^{-\sigma} \sin^2\theta\right]$$

$$= \frac{1}{32}\int_0^\pi d\theta \sin^3\theta \int_0^\infty dr\ r^2\sigma^2 e^{-\sigma}$$

$$= \frac{1}{32}\int_{-1}^{1} dx\ (1 - x^2) \int_0^\infty d\sigma\ \sigma^4 e^{-\sigma} = \frac{1}{32}\left[\frac{4}{3}\right]\left[4!\right] = 1$$

We now show that the first few hydrogen atomic wave functions given in Table 6-5 are orthogonal.

$$\int_0^{2\pi} d\phi \int_0^\pi d\theta \sin\theta \int_0^\infty dr\ r^2 \left[\frac{1}{\pi^{1/2}}\left[\frac{Z}{a_0}\right]^{3/2} e^{-\sigma}\right]$$

$$\times \left[\frac{1}{\sqrt{32\pi}}\left[\frac{Z}{a_0}\right]^{3/2} (2 - \sigma) e^{-\sigma/2}\right]$$

$$= \frac{1}{\sqrt{2}}\left[\frac{Z}{a_0}\right]^3 \int_0^\infty dr\ r^2 (2 - \sigma) e^{-3\sigma/2}$$

$$= \frac{1}{\sqrt{2}}\int_0^\infty d\sigma (2\sigma^2 - \sigma^3) e^{-3\sigma/2}$$

$$= \frac{1}{\sqrt{2}}\left[2\cdot 2!\left[\frac{2}{3}\right]^3 - 3!\left[\frac{2}{3}\right]^4\right] = 0$$

In showing that either $\psi_{100}$ or $\psi_{200}$ is orthogonal to $\psi_{210}$, the integral over $\theta$ is

$$\int_0^\pi d\theta \sin\theta \cos\theta = \int_{-1}^{1} dx\ x = 0$$

In showing that either $\psi_{100}$, $\psi_{200}$ or $\psi_{210}$ is orthogonal to $\psi_{21\pm1}$, the integral over $\phi$ is

$$\int_0^{2\pi} d\phi e^{\pm i\phi} = 0$$

**6-29.** This problem is similar to Example 6-10. The 1s hydrogen atomic wave function is (Table 6-5)

$$\psi_{100} = \frac{1}{\sqrt{\pi}}\left[\frac{Z}{a_0}\right]^{3/2} e^{-\sigma}$$

The probability that a hydrogen 1s electron will be found within a distance $2a_0$ of the nucleus is

$$\text{prob} = \int_0^{2\pi} d\phi \int_0^{\pi} d\theta \sin\theta \int_0^{2a_0} dr\ r^2 \left[\frac{1}{\pi}\left[\frac{Z}{a_0}\right]^3 e^{-2\sigma}\right]$$

$$= 4\left[\frac{Z}{a_0}\right]^3 \int_0^{2a_0} dr\ r^2 e^{-2\sigma}$$

$$= 4\int_0^{2} d\sigma\ \sigma^2 e^{-2\sigma} = 4\left[\frac{1}{4} - \frac{13}{4}e^{-4}\right]$$

$$= 1 - 13e^{-4} = 0.762$$

**6-30.** From Table 6-5,

$$\psi_{100} = \frac{1}{\sqrt{\pi}}\left[\frac{Z}{a_0}\right]^3 e^{-\sigma}$$

The probability that a hydrogen 1s electron will be found within a distance $Da_0$ of the nucleus is given by

$$\text{prob}(D) = \int_0^{2\pi} d\phi \int_0^{\pi} d\theta \sin\theta \int_0^{Da_0} dr\ r^2 \left[\frac{1}{\pi}\left[\frac{Z}{a_0}\right]^3 e^{-2\sigma}\right]$$

$$= 4\int_0^{D} d\sigma\ \sigma^2 e^{-2\sigma} = 1 - e^{-2D}(2D^2 + 2D + 1)$$

Now let prob(D) = 0.500 and solve for D by trial and error.

$$0.500 = 1 - e^{-2D}(2D^2 + 2D + 1)$$

to get D = 1.34. Similarly, let prob(D) = 0.900 and find by trial and error that D = 2.66.

**6-31.** We start with

$$I(\beta) = \int_0^{\infty} e^{-\beta r} dr = \frac{1}{\beta}$$

The derivatives of $I(\beta)$ are

$$\frac{dI}{d\beta} = \int_0^{\infty} \frac{\partial}{\partial\beta}(e^{-\beta r}) dr = -\int_0^{\infty} r e^{-\beta r} dr$$

$$\frac{d^2 I}{d\beta^2} = \int_0^{\infty} \frac{\partial^2}{\partial\beta^2}(e^{-\beta r}) dr = \int_0^{\infty} r^2 e^{-\beta r} dr$$

and so on. Because $I(\beta) = 1/\beta$, we have that

$$\frac{dI}{d\beta} = -\frac{1}{\beta^2} \qquad\qquad \frac{d^2I}{d\beta^2} = \frac{2}{\beta^3}$$

Generally,

$$\frac{d^nI}{d\beta^n} = \int_0^\infty \frac{\partial^n}{\partial\beta^n}(e^{-\beta r})dr = (-1)^n\int_0^\infty r^n e^{-\beta r}dr$$

$$= \frac{d^n}{d\beta^n}\beta^{-1} = (-1)^n \frac{n!}{\beta^{n+1}}$$

Therefore,

$$\frac{d^nI}{d\beta^n} = \frac{n!}{\beta^{n+1}}$$

**6-32.**

$$\langle r\rangle = \int_0^{2\pi} d\phi \int_0^{\pi} d\theta\sin\theta \int_0^\infty dr\; r^2\psi^*_{n\ell m}(r,\theta,\phi)r\psi_{n\ell m}(r,\theta,\phi)$$

For the 1s state

$$\langle r\rangle_{1s} = 4\left[\frac{Z}{a_0}\right]^3 \int_0^\infty dr\; r^3e^{-2\sigma}$$

$$= \frac{4a_0}{Z}\int_0^\infty d\sigma\; \sigma^3e^{-2\sigma} = \frac{4a_0}{Z}\left[\frac{3!}{16}\right] = \frac{3}{2}\,\frac{a_0}{Z}$$

For the 2s state

$$\langle r\rangle_{2s} = \frac{1}{8}\left[\frac{Z}{a_0}\right]^3 \int_0^\infty dr\; r^3(4 - 4\sigma + \sigma^2)e^{-\sigma}$$

$$= \frac{a_0}{8Z}\int_0^\infty d\sigma(4\sigma^3 - 4\sigma^4 + \sigma^5)e^{-\sigma}$$

$$= \frac{a_0}{8Z}(4\cdot3! - 4\cdot4! + 5!) = \frac{6a_0}{Z}$$

**6-33.** The average potential energy of a hydrogen-like 2s electron is given by

$$\langle U\rangle = \left\langle -\frac{e^2}{4\pi\varepsilon_0 r}\right\rangle$$

$$= \int_0^{2\pi} d\phi \int_0^{\pi} d\theta \sin\theta \int_0^{\infty} dr\ r^2 \psi_{2s}^* \left[-\frac{e^2}{4\pi\varepsilon_0 r}\right] \psi_{2s}$$

$$= -\frac{e^2}{32\pi\varepsilon_0}\left[\frac{Z}{a_0}\right]^3 \int_0^{\infty} dr\ r^2 \left[\frac{1}{r}\right](2-\sigma)^2 e^{-\sigma}$$

$$= -\frac{e^2}{32\pi\varepsilon_0}\frac{Z}{a_0}\int_0^{\infty} d\sigma(4\sigma - 4\sigma^2 + \sigma^3)e^{-\sigma}$$

$$= -\frac{e^2}{16\pi\varepsilon_0}\frac{Z}{a_0}$$

The total energy is given by

$$E = -\frac{e^2 Z}{32\pi\varepsilon_0 a_0}$$

and so we see that

$$\langle U\rangle = 2\langle E\rangle$$

Using the fact that $\langle K\rangle + \langle U\rangle = E$, we also have that

$$\langle K\rangle = -E$$

**6-34.** The average value of r in a hydrogen 2s orbital is shown in Problem 6-32 to be

$$\langle r\rangle_{2s} = 6a_0$$

For a 2p state we write

$$\langle r\rangle_{2p} = \int_0^{2\pi} d\phi \int_0^{\pi} d\theta \sin\theta \int_0^{\infty} dr\ r^3 \left[\frac{1}{32\pi}\left[\frac{1}{a_0}\right]^3 \sigma^2 e^{-\sigma}\cos^2\theta\right]$$

$$= \frac{1}{16a_0^3}\int_0^{\pi} d\theta \sin\theta\cos^2\theta \int_0^{\infty} dr\ r^3\sigma^2 e^{-\sigma}$$

$$= \frac{a_0}{16}\int_{-1}^{1} dx\ x^2 \int_0^{\infty} d\sigma\ \sigma^5 e^{-\sigma} = \frac{a_0}{16}\left[\frac{2}{3}\right](5!) = 5a_0$$

and for a 3s state

$$\langle r\rangle_{3s} = \int_0^{2\pi} d\phi \int_0^{\pi} d\theta \sin\theta$$

$$\times \int_0^{\infty} dr\ r^3 \left[\frac{1}{3^9\pi a_0^3}(27 - 18\sigma + 2\sigma^2)^2 e^{-2\sigma/3}\right]$$

$$= \frac{4a_0}{3^9}\int_0^\infty d\sigma\ \sigma^3[3^6 + 4\cdot 3^4\sigma^2 + 4\sigma^4 - 4\cdot 3^5\sigma + 4\cdot 3^3\sigma^2 - 8\cdot 3^2\sigma^3]e^{-2\sigma/3}$$

$$= \frac{4a_0}{3^9}\left[3^6\cdot 3!\left[\frac{3}{2}\right]^4 + 4\cdot 3^4\cdot 5!\left[\frac{3}{2}\right]^6 + 4\cdot 7!\left[\frac{3}{2}\right]^8 - 4\cdot 3^5\cdot 4!\left[\frac{3}{2}\right]^5 + 4\cdot 3^3\cdot 5!\left[\frac{3}{2}\right]^6 - 8\cdot 3^2\cdot 6!\left[\frac{3}{2}\right]^7\right]$$

$$= \frac{27a_0}{2}$$

All these results agree with the general formula that is given in the problem.

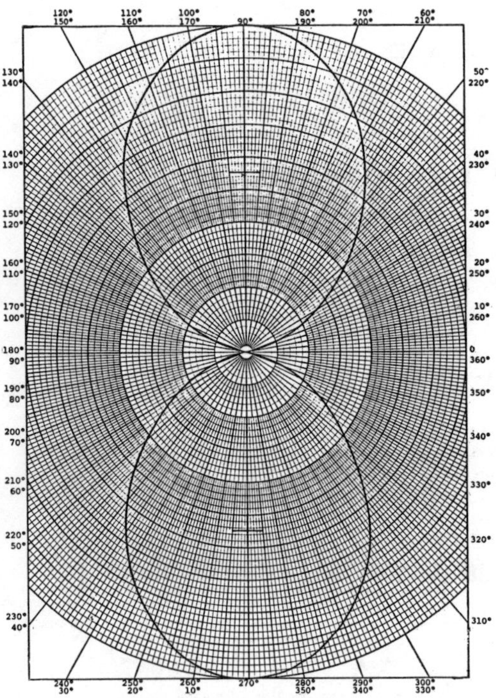

$f(\theta) = \sin^2\theta$

$f(\theta) = \frac{1}{2}(5\cos^3\theta - 3\cos\theta)$

**6-36.** To verify the contour lines in Figure 6-13a, we use the result derived in Problem 6-30

$$\text{prob}(D) = 1 - e^{-D}(2D^2 + 2D + 1)$$

or

$$e^{-2D}(2D^2 + 2D + 1) = 1 - \text{prob}(D)$$

where prob(D) is the probability of finding the electron within D Bohr radii of the nucleus. The value of D must be calculated by trial and error, and the results are

| prob(D) | D |
|---|---|
| 0.100 | 0.551 |
| 0.200 | 0.768 |
| 0.300 | 0.957 |
| 0.400 | 1.14 |
| 0.500 | 1.34 |
| 0.600 | 1.55 |
| 0.700 | 1.81 |
| 0.800 | 2.14 |
| 0.900 | 2.66 |

**6-37.** This problem is very similar to Problem 6-28.

**6-38.** The three transitions are shown in Figure 6-16. The frequencies of the transitions are given by (see page 241)

$$\nu = \frac{\Delta E}{h} = \frac{(1.63\times10^{-18} + 9.27\times10^{-23}\ m)\text{J}}{h} \qquad m = 0,\ \pm1$$

$$= ((2.46 \times 10^{15}) + (1.40 \times 10^{10}\ m))\text{s}^{-1}$$

**6-39.** The Hamiltonian operator for a hydrogen-like system is

$$-\frac{\hbar^2}{2\mu}\left[\frac{1}{r^2}\frac{\partial}{\partial r}\left[r^2\frac{\partial}{\partial r}\right] + \frac{1}{r^2\sin\theta}\frac{\partial}{\partial\theta}\left[\sin\theta\frac{\partial}{\partial\theta}\right] + \frac{1}{r^2\sin^2\theta}\frac{\partial^2}{\partial\phi^2}\right] - \frac{Ze^2}{4\pi\varepsilon_0 r}$$

The only difference between this Hamiltonian and that of a hydrogen atom is that $e^2$ is replaced by $Ze^2$.

Because e appears nowhere else, we can obtain the results of this problem by replacing $e^2$ by $Ze^2$ in the hydrogen atom results.

For example, the expression for the energy of a hydrogen atom is (Eq. 6-99)

$$E_n = -\frac{\mu e^4}{8\varepsilon_0{}^2h^2n^2} \qquad \text{(hydrogen atom)}$$

and so the energy of a hydrogen-like system is given by

$$E_n = -\frac{\mu Z^2e^4}{8\varepsilon_0{}^2h^2n^2} \qquad \text{(hydrogen-like system)}$$

Furthermore, the Bohr radius ($a_0 = \varepsilon_0h^2/\pi\mu e^2$) becomes

$$a_0 = \frac{\varepsilon_0h^2}{\pi\mu Ze^2}$$

Consequently, we replace $a_0$ by $a_0/Z$ in Eq. 6-102

$$R_{n\ell}(r) = -\left[\frac{(n-\ell-1)!}{2n[(n+\ell)!]^3}\right]^{1/2} \times \left[\frac{2Z}{na_0}\right]^{\ell+(3/2)} r^{\ell}\exp(-Zr/na_0)L_{n+\ell}^{2\ell+1}\left[\frac{2Zr}{na_0}\right]$$

The hydrogen atomic 1s orbital

$$\psi_{1s} = \frac{1}{\pi^{1/2}a_0{}^{3/2}}\exp(-r/a_0) \qquad \text{(hydrogen atom)}$$

becomes

$$\psi_{1s} = \frac{1}{\sqrt{\pi}}\left[\frac{Z}{a_0}\right]^{3/2}\exp(-Zr/a_0) \qquad \text{(hydrogen-like system)}$$

Using this result

$$\langle r\rangle_{1s} = \int_0^{2\pi}d\phi\int_0^{\pi}d\theta\sin\theta\int_0^{\infty}dr\ r^3\left[\frac{Z^3}{\pi a_0{}^3}\exp(-2Zr/a_0)\right]$$

$$= \frac{4Z^3}{a_0{}^3}\int_0^{\infty}dr\ r^3\exp(-2Zr/a_0) = \frac{4a_0}{Z}\int_0^{\infty}d\sigma\ \sigma^3e^{-2\sigma}$$

$$= \frac{4a_0}{Z}\ \frac{3!}{16} = \frac{3a_0}{2Z}$$

and $r_{mp}$ is given by

$$\frac{d}{dr}(r^2\exp(-2Zr/a_0)) = 0$$

or $r = a_0/Z$.

The ionizaton energy is given by $-E_1$, and so (see Example 1-9)

$$IE = \frac{\mu Z^2 e^4}{8\varepsilon_0^2 h^2} = (13.6 \text{ eV})Z^2$$

Therefore

$$IE_H = 13.6 \text{ eV}$$

and

$$IE_{He^+} = 54.4 \text{ eV}$$

**6-40.** We first prove that

$$\int \psi^* [\hat{H}, \hat{A}] \psi d\tau = 0$$

Writing out the commutator, we have

$$\int \psi^* \hat{H}\hat{A}\psi d\tau - \int \psi^* \hat{A}\hat{H}\psi d\tau \stackrel{?}{=} 0$$

Using the Hermitian property of $\hat{H}$, we can write

$$\int (\hat{H}\psi)^* \hat{A}\psi d\tau - \int \psi^* \hat{A}(\hat{H}\psi) d\tau \stackrel{?}{=} 0$$

But

$$\hat{H}\psi = E\psi$$

where E is real, and so we have

$$E\int \psi^* \hat{A}\psi d\tau - E\int \psi^* \hat{A}\psi d\tau = 0$$

If

$$\hat{A} = -i\hbar\left[x\frac{\partial}{\partial x} + y\frac{\partial}{\partial y} + z\frac{\partial}{\partial z}\right]$$

then

$$\hat{H}\hat{A}f = -i\hbar\left[-\frac{\hbar^2}{2m}\nabla^2 + U(x,y,z)\right]\left[x\frac{\partial f}{\partial x} + y\frac{\partial f}{\partial y} + z\frac{\partial f}{\partial z}\right]$$

$$= \frac{i\hbar^3}{2m}\left[2\frac{\partial^2 f}{\partial x^2} + x\frac{\partial^3 f}{\partial x^3} + y\frac{\partial^3 f}{\partial x^2 \partial y} + z\frac{\partial^3 f}{\partial x^2 \partial z}\right.$$

$$+ x\frac{\partial^3 f}{\partial x \partial y^2} + 2\frac{\partial^2 f}{\partial y^2} + y\frac{\partial^3 f}{\partial y^3} + z\frac{\partial^3 f}{\partial y^2 \partial z}$$

$$\left. + x\frac{\partial^3 f}{\partial x \partial z^2} + y\frac{\partial^3 f}{\partial y \partial z^2} + 2\frac{\partial^2 f}{\partial z^2} + z\frac{\partial^3 f}{\partial z^3}\right]$$

$$- i\hbar\left[U(x,y,z)\left[x\frac{\partial f}{\partial x} + y\frac{\partial f}{\partial y} + z\frac{\partial f}{\partial z}\right]\right]$$

and

$$\hat{A}\hat{H}f = \frac{i\hbar^3}{2m}\left[x\frac{\partial f}{\partial x} + y\frac{\partial f}{\partial y} + z\frac{\partial f}{\partial z}\right]\left[\frac{\partial^2 f}{\partial x^2} + \frac{\partial^2 f}{\partial y^2} + \frac{\partial^2 f}{\partial z^2}\right]$$

$$-i\hbar\left[x\frac{\partial}{\partial x} + y\frac{\partial}{\partial y} + z\frac{\partial}{\partial z}\right]U(x,y,z)f(x,y,z)$$

$$= \frac{i\hbar^3}{2m}\left[x\frac{\partial^3 f}{\partial x^3} + x\frac{\partial^3 f}{\partial x\,\partial y^2} + x\frac{\partial^3 f}{\partial x\,\partial z^2}\right.$$

$$+ y\frac{\partial^3 f}{\partial x^2\partial y} + y\frac{\partial^3 f}{\partial y^3} + y\frac{\partial^3 f}{\partial y\,\partial z^2}$$

$$\left.+ z\frac{\partial^3 f}{\partial x^2\partial z} + z\frac{\partial^3 f}{\partial y^2\partial z} + z\frac{\partial^3 f}{\partial z^3}\right]$$

$$- i\hbar\left[\left[x\frac{\partial U}{\partial x} + y\frac{\partial U}{\partial y} + z\frac{\partial U}{\partial z}\right]f(x,y,z)\right.$$

$$\left.+ \left[x\frac{\partial f}{\partial x} + y\frac{\partial f}{\partial y} + z\frac{\partial f}{\partial z}\right]U(x,y,z)\right]$$

Therefore

$$(\hat{H}\hat{A} - \hat{A}\hat{H})f = \frac{i\hbar^3}{2m}\left[2\frac{\partial^2 f}{\partial x^2} + 2\frac{\partial^2 f}{\partial y^2} + 2\frac{\partial^2 f}{\partial z^2}\right]$$

$$+ i\hbar\left[x\frac{\partial U}{\partial x} + y\frac{\partial U}{\partial y} + z\frac{\partial U}{\partial z}\right]f$$

or

$$\hat{H}\hat{A} - \hat{A}\hat{H} = i\hbar\left[x\frac{\partial U}{\partial x} + y\frac{\partial U}{\partial y} + z\frac{\partial U}{\partial z}\right]$$

$$- \frac{i\hbar}{m}(\hat{P}_x{}^2 + \hat{P}_y{}^2 + \hat{P}_z{}^2)$$

$$= i\hbar\left[x\frac{\partial U}{\partial x} + y\frac{\partial U}{\partial y} + z\frac{\partial U}{\partial z}\right] - 2i\hbar\hat{K}$$

where

$$\hat{K} = \frac{1}{2m}(\hat{P}_x{}^2 + \hat{P}_y{}^2 + \hat{P}_z{}^2)$$

Given that

$$\int\psi^*[\hat{H},\hat{A}]\psi d\tau = 0$$

we can write

$$\left\langle x\frac{\partial U}{\partial x} + y\frac{\partial U}{\partial y} + z\frac{\partial U}{\partial z}\right\rangle = 2\langle K\rangle$$

If

$$U(x,y,z) = \frac{Ze^2}{4\pi\varepsilon_0(x^2 + y^2 + z^2)^{1/2}}$$

then

$$x\frac{\partial U}{\partial x} + y\frac{\partial U}{\partial y} + z\frac{\partial U}{\partial z} = -\frac{Ze^2}{4\pi\varepsilon_0}\left[\frac{x^2}{r^3} + \frac{y^2}{r^3} + \frac{z^2}{r^3}\right]$$

$$= \frac{Ze^2}{4\pi\varepsilon_0 r} = -U$$

and so

$$\langle U\rangle = -2\langle K\rangle$$

Since

$$\langle K\rangle + \langle U\rangle = E$$

We can also write

$$\langle U\rangle = E - \langle K\rangle = -2\langle K\rangle$$

or

$$E = -\langle K\rangle$$

**6-41.** For a harmonic oscillator

$$U(x,y,z) = \frac{k_x x^2}{2} + \frac{k_y y^2}{2} + \frac{k_z z^2}{2}$$

Therefore,

$$x\frac{\partial U}{\partial x} + y\frac{\partial U}{\partial y} + z\frac{\partial U}{\partial z} = k_x x^2 + k_y y^2 + k_z z^2 = 2U$$

and

$$2\langle U\rangle = 2\langle K\rangle$$

Since

$$\langle K\rangle + \langle U\rangle = E$$

we can also write

$$\langle K\rangle = \langle U\rangle = E/2$$

# CHAPTER SEVEN

## APPROXIMATE METHODS

**7-1.** a) The first term in $U(x)$ is $(1/2)kx^2$, the harmonic oscillator potential, and so we shall take a harmonic oscillator as the reference system. Thus

$$\hat{H}^{(0)} = -\frac{\hbar^2}{2\mu}\frac{d^2}{dx^2} + \frac{1}{2}kx^2 \qquad (5\text{-}31)$$

$$\psi_n^{(0)}(x) = N_n H_n(\alpha^{1/2}x)e^{-\alpha x^2/2} \qquad (5\text{-}41)$$

$$E_n^{(0)} = h\nu(n + 1/2) \qquad (5\text{-}33)$$

and

$$\hat{H}^{(1)} = \frac{\gamma}{6}x^3 + \frac{b}{24}x^4$$

b) The particle is constrained to move in the region $0 \le x \le a$ and so we shall take a particle in a box as the reference system. Thus

$$\hat{H}^{(0)} = -\frac{\hbar^2}{2m}\frac{d^2}{dx^2} \qquad 0 \le x \le a$$

$$\psi_n^{(0)}(x) = \left[\frac{2}{a}\right]^{1/2}\sin\frac{n\pi x}{a} \qquad 0 \le x \le a$$

$$E_n^{(0)} = \frac{n^2h^2}{8ma^2}$$

and

$$\hat{H}^{(1)} = 0 \qquad 0 < x < a/2$$

$$= b \qquad a/2 < x < a$$

c) If we neglect the interelectronic repulsion term in the helium atom Hamiltonian, then we have

$$\hat{H}_{He}(0) = \hat{H}_H(1) + \hat{H}_H(2)$$

where $\hat{H}_H(j)$ is the hydrogen-like Hamiltonian operator. Using the results of Section 6-2 then, we have

$$\psi^{(0)}(\mathbf{r}_1,\mathbf{r}_2) = \psi_{n\ell m}(\mathbf{r}_1)\psi_{n'\ell'm'}(\mathbf{r}_2)$$

where $\psi_{n\ell m}(\mathbf{r}_j)$ are given in Table 6-5 or Eq. 6-103.

The unperturbed energy is (Eq. 6-99)

$$E^{(0)} = -\frac{\mu e^4}{8\varepsilon_0^2 h^2 n_1^2} - \frac{\mu e^4}{8\varepsilon_0^2 h^2 n_2^2}$$

and $\hat{H}^{(1)}$ is

$$\hat{H}^{(1)} = \frac{e^2}{4\pi\varepsilon_0 r_{12}}$$

d) We take a hydrogen atom in the absence of an electric field as the reference system, so that

$$\hat{H}^{(0)} = -\frac{\hbar^2}{2\mu}\nabla^2 - \frac{e^2}{4\pi\varepsilon_0 r}$$

$$\psi^{(0)} = \psi_{n\ell m}(r, \theta, \phi) \qquad (6\text{-}103)$$

$$E^{(0)} = -\frac{\mu e^4}{8\varepsilon_0^2 h^2 n^2} \qquad (6\text{-}99)$$

and

$$\hat{H}^{(1)} = e\mathcal{E}r\cos\theta$$

e) We take a rigid rotator in the absence of an electric field as the reference system, so that

$$\hat{H}^{(0)} = -\frac{\hbar^2}{2I}\nabla^2$$

$$\psi^{(0)} = Y_\ell^m(\theta, \phi) \qquad (6\text{-}76)$$

$$E^{(0)} = \frac{\hbar^2}{2I}\ell(\ell + 1) \qquad (6\text{-}61)$$

and

$$\hat{H}^{(1)} = \mu\mathcal{E}\cos\theta$$

**7-2.** Use Eq. 7-20,

$$\Delta E = \int \psi^{(0)*}\hat{H}^{(1)}\psi^{(0)}d\tau$$

$$= \int_{-\infty}^{\infty} dx\, \psi_1^*(x)\left[\frac{\gamma}{6}x^3 + \frac{b}{24}x^4\right]\psi_1(x)$$

$$= 2\alpha\left[\frac{\alpha}{\pi}\right]^{1/2}\int_{-\infty}^{\infty} dx\; x^2 e^{-\alpha x^2}\left[\frac{\gamma}{6}x^3 + \frac{b}{24}x^4\right]$$

$$= 0 + \frac{\alpha b}{6}\left[\frac{\alpha}{\pi}\right]^{1/2}\int_0^\infty dx\ x^6 e^{-\alpha x^2}$$

$$= \frac{5b}{32\alpha^2}$$

**7-3.** $\Delta E = \int \psi^{(0)*}\hat{H}^{(1)}\psi^{(0)}d\tau$

$$= \frac{2b}{a}\int_{a/2}^{a} \sin^2 \frac{\pi x}{a}dx$$

$$= \frac{2b}{a}\left|\frac{x}{2} - \frac{a}{4\pi}\sin\frac{2\pi x}{a}\right|_{a/2}^{a} = \frac{2b}{a}\left[\frac{a}{2} - \frac{a}{4}\right] = \frac{b}{2}$$

**7-4.** We start with

$$\Delta E = e\mathcal{E}\int d\mathbf{r}\psi^*_{n\ell m}(\mathbf{r})r\cos\theta\psi_{n\ell m}(\mathbf{r})$$

Using Eq. 6-103 for the hydrogen atomic wave functions,

$$\Delta E = e\mathcal{E}\int_0^\infty dr\ r^3R_{10}^2(r)\int_0^{2\pi}d\phi\int_0^\pi d\theta\ \sin\theta\ Y_0^0(\theta,\phi)$$
$$\times \cos\theta\ Y_0^0(\theta,\phi)$$

The integral over $\theta$ is

$$\int_0^\pi d\theta\sin\theta\cos\theta\left[\frac{1}{4\pi}\right] = \frac{1}{4\pi}\int_{-1}^{1}dx\ x = 0$$

and so

$$\Delta E = 0$$

**7-5.**

$$\Delta E = \frac{2}{a}\int_0^a U(x)\sin^2 \frac{n\pi x}{a}dx$$

$$= \frac{2V}{a}\int_0^{a/2}x\sin^2 \frac{n\pi x}{a}dx + \frac{2V}{a}\int_{a/2}^{a}(a - x)\sin^2 \frac{n\pi x}{a}dx$$

$$= \frac{2V}{a}\left|\frac{x^2}{4} - \frac{ax\sin(2\pi nx/a)}{4n\pi} - \frac{a^2\cos(2\pi nx/a)}{8n^2\pi^2}\right|_0^{a/2}$$

$$+ 2V\left|_{a/2}^{a} \frac{x}{2} - \frac{a}{4n\pi}\sin\frac{2\pi nx}{a}\right|$$

$$- \frac{2V}{a}\left|_{a/2}^{a} \frac{x^2}{4} - \frac{ax\sin(2\pi nx/a)}{4n\pi} - \frac{a^2\cos(2\pi nx/a)}{8n^2\pi^2}\right|$$

$$= \frac{2V}{a}\left[\frac{a^2}{16} - \frac{a^2\cos n\pi}{8n^2\pi^2} + \frac{a^2}{8n^2\pi^2}\right]$$

$$+ 2V\left[\frac{a}{2} - \frac{a}{4}\right] - \frac{2V}{a}\left[\frac{a^2}{4} - \frac{a^2}{8n^2\pi^2} - \frac{a^2}{16} + \frac{a^2\cos n\pi}{8n^2\pi^2}\right]$$

$$= \frac{Va}{2}\left[\frac{1}{2} + \frac{(1 - \cos n\pi)}{n^2\pi^2}\right]$$

**7-6.**

$$\Delta E = \frac{8V}{abc}\int_{a/4}^{3a/4} dx \int_{b/4}^{3b/4} dy \int_{c/4}^{3c/4} dz \, \sin^2 \frac{\pi x}{a}\sin^2 \frac{\pi y}{b}\sin^2 \frac{\pi z}{c}$$

$$= \frac{8V}{abc}\left[\int_{a/4}^{3a/4} dx\sin^2 \frac{\pi x}{a}\right]\left[\int_{b/4}^{3b/4} dy\sin^2 \frac{\pi x}{b}\right]\left[\int_{c/4}^{3c/4} dz\sin^2 \frac{\pi x}{c}\right]$$

A typical integral here is

$$\int_{a/4}^{3a/4} dx\sin^2 \frac{\pi x}{a} = \left|_{a/4}^{3a/4} \frac{x}{2} - \frac{a}{4\pi}\sin\frac{2\pi x}{a}\right|$$

$$= \frac{a}{4} + \frac{a}{2\pi}$$

Therefore

$$\Delta E = \frac{8V}{abc}\left[\frac{a}{4} + \frac{a}{2\pi}\right]\left[\frac{b}{4} + \frac{b}{2\pi}\right]\left[\frac{c}{4} + \frac{c}{2\pi}\right]$$

$$= 8V\left[\frac{1}{4} + \frac{1}{2\pi}\right]^3 = 0.548 \text{ V}$$

**7-7.** The Hamiltonian operator is

$$\hat{H} = -\frac{\hbar^2}{2\mu}\frac{d^2}{dx^2} + cx^4$$

In order to use a harmonic oscillator as the reference system, add and subtract $(1/2)kx^2$ to $\hat{H}$ to obtain

$$\hat{H} = -\frac{\hbar^2}{2\mu}\frac{d^2}{dx^2} + \frac{1}{2}kx^2 + cx^4 - \frac{1}{2}kx^2$$

Therefore

$$\hat{H}^{(1)} = cx^4 - \frac{1}{2}kx^2$$

and

$$\Delta E = \left[\frac{\alpha}{\pi}\right]^{1/2}\int_{-\infty}^{\infty} dx e^{-\alpha x^2}\left[cx^4 - \frac{1}{2}kx^2\right]$$

$$= \left[\frac{\alpha}{\pi}\right]^{1/2} 2\left[\frac{3c}{8\alpha^2}\left[\frac{\pi}{\alpha}\right]^{1/2} - \frac{k}{8\alpha}\left[\frac{\pi}{\alpha}\right]^{1/2}\right]$$

$$= \frac{3c}{4\alpha^2} - \frac{k}{4\alpha}$$

**7-8.**

$$\Delta E = \frac{2}{\ell}\int_0^{\ell} U(x)\sin^2 \frac{n\pi x}{\ell}dx$$

$$= \frac{2A}{\ell}\int_0^{\ell}\left[x^2 - x\ell + \frac{\ell^2}{4}\right]\sin^2 \frac{n\pi x}{\ell}dx$$

$$= A\left[\langle x^2\rangle - \ell\langle x\rangle + \frac{\ell^2}{4}\right]$$

Using

$$\langle x^2\rangle = \left[\frac{\ell}{2\pi n}\right]^2\left[\frac{4\pi^2 n^2}{3} - 2\right] \qquad (3\text{-}46)$$

and

$$\langle x\rangle = \frac{\ell}{2} \qquad (3\text{-}45)$$

we have

$$\Delta E = A\left[\frac{\ell^2}{12} - \frac{\ell^2}{2\pi^2 n^2}\right] = \frac{A\ell^2}{12}\left[1 - \frac{6}{\pi^2 n^2}\right]$$

**7-9.** Let $\phi = \psi + \varepsilon\eta$, where $\psi$ is the exact ground-state wave function. Then

$$E_\phi = \int\phi^*\hat{H}\phi d\tau = \int(\psi + \varepsilon\eta)^*\hat{H}(\psi + \varepsilon\eta)d\tau$$

$$= \int\psi^*\hat{H}\psi d\tau + \varepsilon^*\int\eta^*\hat{H}\psi d\tau + \varepsilon\int\psi^*\hat{H}\eta d\tau + O(\varepsilon^2)$$

$$= E_0 + \varepsilon^*\int\eta^*\hat{H}\psi d\tau + \varepsilon\int\psi^*\hat{H}\eta d\tau + O(\varepsilon^2)$$

Thus, $E_\phi - E_0$ is of order $\varepsilon$.

**7-10.** For the ground-state of a hydrogen atom

$$\hat{H} = -\frac{\hbar^2}{2\mu r^2}\frac{d}{dr}\left[r^2\frac{d}{dr}\right] - \frac{e^2}{4\pi\varepsilon_0 r}$$

Therefore

$$E(\alpha) = \frac{\int_0^\infty 4\pi r^2 dr e^{-\alpha r^2}\hat{H}e^{-\alpha r^2}}{\int_0^\infty 4\pi r^2 dr e^{-2\alpha r^2}}$$

$$= \frac{-\frac{\hbar^2}{2\mu}\int_0^\infty dr[4\alpha^2 r^4 - 6\alpha r^2]e^{-2\alpha r^2} - \frac{e^2}{4\pi\varepsilon_0}\int_0^\infty dr\ re^{-2\alpha r^2}}{\int_0^\infty dr\ r^2 e^{-2\alpha r^2}}$$

$$= \frac{-\frac{\hbar^2}{2\mu}\left[-\frac{3}{8}\left[\frac{\pi}{2\alpha}\right]^{1/2}\right] - \frac{e^2}{4\pi\varepsilon_0}\frac{1}{4\alpha}}{\frac{1}{8\alpha}\left[\frac{\pi}{2\alpha}\right]^{1/2}}$$

$$= \frac{3\hbar^2\alpha}{2\mu} - \frac{e^2\alpha^{1/2}}{2^{1/2}\varepsilon_0\pi^{3/2}}$$

The minimum value of $E(\alpha)$ is found from

$$\frac{dE}{d\alpha} = \frac{3\hbar^2}{2\mu} - \frac{e^2}{(2\pi)^{3/2}\varepsilon_0\alpha_{min}^{1/2}} = 0$$

or

$$\alpha_{min} = \frac{\mu^2 e^4}{18\pi^3\varepsilon_0^2\hbar^4}$$

The value $E_{min}$ is

$$E_{min} = E(\alpha_{min}) = -\frac{\mu e^4}{12\pi^3\varepsilon_0^2\hbar^2} = -\frac{4}{3\pi}\left[\frac{\mu e^4}{16\pi^2\varepsilon_0^2\hbar^2}\right]$$

**7-11.** For the ground state of a hydrogen atom

$$\hat{H} = -\frac{\hbar^2}{2\mu r^2}\frac{d}{dr}\left[r^2\frac{d}{dr}\right] - \frac{e^2}{4\pi\varepsilon_0 r}$$

Therefore

$$E(\alpha) = \frac{\int_0^\infty 4\pi r^2 dr e^{-\alpha r}\hat{H}e^{-\alpha r}}{\int_0^\infty 4\pi r^2 dr e^{-2\alpha r}}$$

$$= \frac{-\frac{\hbar^2}{2\mu}\int_0^\infty dr[\alpha^2 r^2 - 2\alpha r]e^{-2\alpha r} - \frac{e^2}{4\pi\varepsilon_0}\int_0^\infty dr\ re^{-2\alpha r}}{\int_0^\infty dr\ r^2 e^{-2\alpha r}}$$

$$= \frac{-\frac{\hbar^2}{2\mu}\left[-\frac{1}{4\alpha}\right] - \frac{e^2}{4\pi\varepsilon_0}\left[\frac{1}{4\alpha^2}\right]}{\frac{1}{4\alpha^3}}$$

$$= \frac{\hbar^2\alpha^2}{2\mu} - \frac{e^2\alpha}{4\pi\varepsilon_0}$$

The minimum value of E is found from

$$\frac{dE}{d\alpha} = 0 = \frac{\hbar^2\alpha_{min}}{\mu} - \frac{e^2}{4\pi\varepsilon_0}$$

or

$$\alpha_{min} = \frac{\mu e^2}{4\pi\varepsilon_0\hbar^2}$$

Therefore

$$E_{min} = E(\alpha_{min}) = -\frac{1}{2}\left[\frac{\mu e^4}{16\pi^2\varepsilon_0^2\hbar^2}\right]$$

which is the exact ground state energy. This occurs because $e^{-\alpha r}$ is of the form of the exact ground-state wave function for atomic hydrogen, and in fact

$$\alpha_{min} = 1/a_0$$

and the trial function turns out to be $e^{-r/a_0}$.

**7-12.** The energy $E_{min}$ will turn out to be the exact ground-state energy of atomic hydrogen because the form of the exact ground-state wave function is contained in the trial function. The parameter $c_2$ will turn out to be zero (making $\beta$ irrelevant) and

$\alpha$ will turn out to be $1/a_0$ (see Problem 7-11). A similar result will occur for the more general trial function.

**7-13.**

$$E(\beta) = \frac{\int_{-\infty}^{\infty} dx\,\phi(x)\hat{H}\phi(x)}{\int_{-\infty}^{\infty} dx\,\phi^2(x)}$$

$$= \frac{\int_{-\infty}^{\infty} dx e^{-\beta x^2}\left[-\frac{\hbar^2}{2\mu}\frac{d^2}{dx^2} + \frac{k}{2}x^2\right]e^{-\beta x^2}}{\int_{-\infty}^{\infty} dx e^{-2\beta x^2}}$$

$$= \frac{-\frac{\hbar^2}{\mu}\int_0^{\infty} dx[4\beta^2x^2 - 2\beta]e^{-2\beta x^2} + k\int_0^{\infty} dx\; x^2 e^{-2\beta x^2}}{2\int_0^{\infty} dx e^{-2\beta x^2}}$$

$$= \frac{\frac{\hbar^2\beta}{2\mu}\left[\frac{\pi}{2\beta}\right]^{1/2} + \frac{k}{8\beta}\left[\frac{\pi}{2\beta}\right]^{1/2}}{\left[\frac{\pi}{2\beta}\right]^{1/2}} = \frac{\hbar^2\beta}{2\mu} + \frac{k}{8\beta}$$

The minimum value of $E(\beta)$ is found from

$$\frac{dE}{d\beta} = 0 = \frac{\hbar^2}{2\mu} - \frac{k}{8\beta^2_{min}}$$

or

$$\beta_{min} = \left[\frac{\mu k}{4\hbar^2}\right]^{1/2}$$

The value of $E_{min}$ is

$$E_{min} = E(\beta_{min}) = \frac{1}{2}\hbar\left[\frac{k}{\mu}\right]^{1/2}$$

which is the exact ground-state energy. This result

is obtained because the form of the trial function is the same as the exact ground-state wave function, which is

$$\psi_0(x) = \left[\frac{\alpha}{\pi}\right]^{1/2} e^{-\alpha x^2/2}$$

where $\alpha = (\mu k/\hbar^2)^{1/2}$, or $\alpha/2 = (\mu k/4\hbar^2)^{1/2}$.

**7–14.** The Hamiltonian of a harmonic oscillator is

$$\hat{H} = -\frac{\hbar}{2\mu}\frac{d^2}{dx^2} + \frac{1}{2}kx^2$$

Using a trial function of the form $\phi(x) = 1/(x^2 + \alpha^2)$, we have

$$E(\alpha) = \frac{\int_{-\infty}^{\infty} dx\, \phi(x)\hat{H}\phi(x)}{\int_{-\infty}^{\infty} dx\, \phi^2(x)}$$

The numerator is

$$N = -\frac{\hbar^2}{2\mu}\int_{-\infty}^{\infty} dx \left[\frac{6x^2 - 2\alpha^2}{(x^2 + \alpha^2)^4}\right] + \frac{k}{2}\int_{-\infty}^{\infty} \frac{dx\ x^2}{(x^2 + \alpha^2)^2}$$

$$= -\frac{\hbar^2}{\mu}\left[6\cdot\frac{\pi}{32\alpha^5} - 2\alpha^2\cdot\frac{5\pi}{32\alpha^7}\right] + k\cdot\frac{\pi}{4\alpha}$$

and the denominator is

$$D = 2\int_0^{\infty}\frac{dx}{(x^2 + \alpha^2)^2} = \frac{\pi}{2\alpha^3}$$

Therefore

$$E(\alpha) = \frac{\dfrac{\pi\hbar^2}{8\mu\alpha^5} + \dfrac{k\pi}{4\alpha}}{\dfrac{\pi}{2\alpha^3}} = \frac{\hbar^2}{4\mu\alpha^2} + \frac{k\alpha^2}{2}$$

The minimum value of E is given by

$$\frac{dE}{d\alpha} = 0 = -\frac{\hbar^2}{2\mu\alpha_{min}^3} + k\alpha_{min}$$

or

$$\alpha_{min} = \left[\frac{\hbar^2}{2\mu k}\right]^{1/4}$$

The value of $E_{min}$ is

$$E_{min} = E(\alpha_{min}) = \frac{1}{\sqrt{2}}\hbar\left[\frac{k}{\mu}\right]^{1/2}$$

**7-15.** For a three-dimensional, spherically symmetric harmonic oscillator,

$$\hat{H} = -\frac{\hbar^2}{2\mu r^2}\frac{d}{dr}\left[r^2\frac{d}{dr}\right] + \frac{k}{2}r^2$$

Using a trial function of the form $e^{-\alpha r^2}$, we have

$$E(\alpha) = \frac{\int_0^\infty 4\pi r^2 dr e^{-\alpha r^2}\hat{H}e^{-\alpha r^2}}{\int_0^\infty 4\pi r^2 e^{-2\alpha r^2}}$$

$$= \frac{-\frac{\hbar^2}{2\mu}\int_0^\infty dr[4\alpha^2 r^4 - 6\alpha r^2]e^{-2\alpha r^2} + \frac{k}{2}\int_0^\infty dr\ r^4 e^{-2\alpha r^2}}{\int_0^\infty dr\ r^2 e^{-2\alpha r^2}}$$

$$= \frac{-\frac{\hbar^2}{2\mu}\left[\frac{3}{8}\left[\frac{\pi}{2\alpha}\right]^{1/2} - \frac{3}{4}\left[\frac{\pi}{2\alpha}\right]^{1/2}\right] + k\frac{3}{64\alpha^2}\left[\frac{\pi}{2\alpha}\right]^{1/2}}{\frac{1}{8\alpha}\left[\frac{\pi}{2\alpha}\right]^{1/2}}$$

$$= \frac{3\hbar^2\alpha}{2\mu} + \frac{3k}{8\alpha}$$

The value of $\alpha$ that minimizes $E(\alpha)$ is given by

$$\frac{dE}{d\alpha} = 0 = \frac{3\hbar^2}{2\mu} - \frac{3k}{8\alpha^2_{min}}$$

or

$$\alpha_{min} = \frac{(\mu k)^{1/2}}{2\hbar}$$

The value of $E_{min} = E(\alpha_{min})$ is

$$E_{min} = \frac{3}{2}\hbar\left[\frac{k}{\mu}\right]^{1/2}$$

This value of $E_{min}$ is the exact result for this system because our trial function is of the same form as the exact wave function, which is

$$\psi(r) = \psi_0(x)\psi_0(y)\psi_0(z)$$

$$= \left[\frac{\alpha}{\pi}\right]^{3/4} \exp[-\alpha(x^2+y^2+z^2)/2] = \left[\frac{\alpha}{\pi}\right]^{3/4} e^{-\alpha r^2/2}$$

with $\alpha = (\mu k)^{1/2}/\hbar$, or $\alpha/2 = (\mu k)^{1/2}/2\hbar$.

Let's go on now and use $e^{-\alpha r}$ as a trial function.

$$E(\alpha) = \frac{-\dfrac{\hbar^2}{2\mu}\displaystyle\int_0^\infty dr\,[\alpha^2 r^2 e^{-2\alpha r} - 2\alpha r e^{-2\alpha r}] + \frac{k}{2}\int_0^\infty dr\; r^4 e^{-2\alpha r}}{\displaystyle\int_0^\infty dr\; r^2 e^{-2\alpha r}}$$

$$= \frac{-\dfrac{\hbar^2}{2\mu}\left[\dfrac{1}{4\alpha} - \dfrac{1}{2\alpha}\right] + \dfrac{k}{2}\left[\dfrac{3}{4\alpha^5}\right]}{\dfrac{1}{4\alpha^3}}$$

$$= \frac{\hbar^2\alpha^2}{2\mu} + \frac{3k}{2\alpha^2}$$

We minimize $E(\alpha)$ with respect to $\alpha$ to find

$$\alpha_{min} = \left[\frac{3\mu k}{\hbar^2}\right]^{1/4}$$

The value of the energy is

$$E_{min} = E(\alpha_{min}) = \sqrt{3}\hbar\left[\frac{k}{\mu}\right]^{1/2}$$

Note that this result is greater than the exact ground-state energy.

**7-16.** We start with

$$E = \frac{\int d\tau \phi^* \hat{H} \phi}{\int d\tau \phi^* \phi}$$

If we substitute $\phi = c_1\phi_1 + c_2\phi_2$ into this expression, then we obtain

$$E = \frac{\int d\tau (c_1^*\phi_1^* + c_2^*\phi_2^*)\hat{H}(c_1\phi_1 + c_2\phi_2)}{\int d\tau (c_1^*\phi_1^* + c_2^*\phi_2^*)(c_1\phi_1 + c_2\phi_2)}$$

$$= \frac{c_1{}^*c_1\int d\tau\phi_1{}^*\hat{H}\phi_1 + c_1{}^*c_2\int d\tau\phi_1{}^*\hat{H}\phi_2 + c_1c_2{}^*\int d\tau\phi_1\hat{H}\phi_2{}^* + c_2{}^*c_2\int d\tau\phi_2{}^*\hat{H}\phi_2}{c_1{}^*c_1\int d\tau\phi_1{}^*\phi_1 + c_1{}^*c_2\int d\tau\phi_1{}^*\phi_2 + c_1c_2{}^*\int d\tau\phi_1\phi_2{}^* + c_2{}^*c_2\int d\tau\phi_2{}^*\phi_2}$$

$$= \frac{c_1{}^*c_1H_{11} + c_1{}^*c_2H_{12} + c_1c_2{}^*H_{21} + c_2{}^*c_2H_{22}}{c_1{}^*c_1S_{11} + c_1{}^*c_2S_{12} + c_1c_2{}^*S_{21} + c_2{}^*c_2S_{21}}$$

Note that

$$H_{ij} = \int\phi_i{}^*\hat{H}\phi_j d\tau = \int\phi_j\hat{H}^*\phi_i{}^*d\tau = H_{ji}{}^*$$

and that

$$S_{ij} = \int\phi_i{}^*\phi_j d\tau = \left[\int\phi_i\,\phi_j{}^*d\tau\right]^* = S_{ji}{}^*$$

Now write

$$c_1{}^*c_1H_{11} + c_1{}^*c_2H_{12} + c_1c_2{}^*H_{21} + c_2{}^*c_2H_{22}$$
$$= E(c_1,c_2,c_1{}^*,c_2{}^*)[c_1{}^*c_1S_{11} + c_1{}^*c_2S_{12} + c_1c_2{}^*S_{21} + c_2{}^*c_2S_{22}]$$

and differentiate with respect to $c_1{}^*$ to obtain

$$c_1H_{11} + c_2H_{12} = \left[\frac{\partial E}{\partial c_1{}^*}\right][c_1{}^*c_1S_{11} + c_1{}^*c_2S_{12} + c_1c_2{}^*S_{21} + c_2{}^*c_2S_{22}] + E(c_1S_{11} + c_2S_{12})$$

If we set $\left[\frac{\partial E}{\partial c_1{}^*}\right] = 0$, then we get

$$c_1(H_{11} - ES_{11}) + c_2(H_{12} - ES_{12}) = 0$$

If we differentiate with respect to $c_2{}^*$ instead of $c_1{}^*$, then the result is

$$c_1(H_{21} - ES_{21}) + c_2(H_{22} - ES_{22}) = 0$$

**7-17.** Our trial function is of the form

$$\phi = c_1\phi_1 + c_2\phi_2$$

where

$$\phi_1 = x(1 - x)$$

and

$$\phi_2 = x^2(1 - x)^2$$

For a particle in a box,

$$\hat{H} = -\frac{\hbar^2}{2m}\frac{d^2}{dx^2}$$

and so

$$H_{11} = -\frac{\hbar^2}{2m}\int_0^1 dx\ x(1-x)\frac{d^2}{dx^2}[x(1-x)]$$

$$= \frac{\hbar^2}{m}\left[\int_0^1 dx\ x(1-x)\right]$$

For this problem, it is convenient to use the result

$$\int_0^1 dx\ x^m(1-x)^n = \frac{m!n!}{(m+n+1)!}$$

Using this to evaluate $H_{11}$ gives

$$H_{11} = \frac{\hbar^2}{m}\left[\frac{(1!)(1!)}{3!}\right] = \frac{\hbar^2}{6m}$$

Similarly

$$H_{12} = H_{21} = -\frac{\hbar^2}{2m}\int_0^1 dx\ x^2(1-x)^2\frac{d^2}{dx^2}[x(1-x)]$$

$$= \frac{\hbar^2}{m}\int_0^1 dx\ x^2(1-x)^2 = \frac{\hbar^2}{m}\left[\frac{(2!)(2!)}{5!}\right]$$

$$= \frac{\hbar^2}{30m}$$

and

$$H_{22} = -\frac{\hbar^2}{2m}\int_0^1 dx\ x^2(1-x)^2\frac{d^2}{dx^2}[x^2(1-x)^2]$$

$$= -\frac{\hbar^2}{2m}\int_0^1 dx\ x^2(1-x)^2[2-12x+12x^2]$$

$$= -\frac{\hbar^2}{2m}\left[\frac{2}{30}-\frac{1}{5}+\frac{4}{35}\right] = \frac{\hbar^2}{105m}$$

$$S_{11} = \int_0^1 dx\ x^2(1-x)^2 = \frac{(2!)(2!)}{5!} = \frac{1}{30}$$

$$S_{12} = S_{21} = \int_0^1 dx\ x^3(1-x)^3 = \frac{(3!)(3!)}{7!} = \frac{1}{140}$$

$$S_{22} = \int_0^1 dx\ x^4(1 - x)^4 = \frac{(4!)(4!)}{9!} = \frac{1}{630}$$

The secular determinantal equation is

$$\begin{vmatrix} \frac{\hbar^2}{6m} - \frac{E}{30} & \frac{\hbar^2}{30m} - \frac{E}{140} \\ \frac{\hbar^2}{30m} - \frac{E}{140} & \frac{\hbar^2}{105m} - \frac{E}{630} \end{vmatrix} = 0$$

Factor out $\hbar^2/m$ from each term and multiply through by 1260 and obtain

$$\begin{vmatrix} 210 - 42E' & 42 - 9E' \\ 42 - 9E' & 12 - 2E' \end{vmatrix} = 0$$

or

$$3E'^2 - 168E' + 756 = 0$$

The two roots to this equation are given by

$$E' = \frac{168 \pm \sqrt{28224 - 9072}}{6}$$

$$= \frac{168 \pm \sqrt{19152}}{6}$$

The smaller root is

$$E' = 4.93487$$

or

$$E = 4.93487\ \hbar^2/m$$

$$= 0.125002\ h^2/m$$

**7-18.** Start with Eq. 7-61. Because of the orthogonality of the $\psi_n$, the $S_{ij} = \delta_{ij}$, and consequently Eq. 7-61 reduces to the secular determinantal equation given in the problem.

**7-19.** We must first evaluate the $H_{ij}$ for this problem. The Hamiltonian for this system is

$$\hat{H} = -\frac{\hbar^2}{2m}\frac{d^2}{dx^2} + Vx \qquad\qquad 0 \le x \le 1$$

Therefore

$$H_{11} = 2\int_0^1 dx\,(\sin\pi x)\hat{H}(\sin\pi x)$$

$$= \frac{\pi^2\hbar^2}{m}\int_0^1 dx \sin^2\pi x + 2V\int_0^1 dx\ x\sin^2\pi x$$

$$= \frac{\pi^2\hbar^2}{m}\left[\frac{1}{2}\right] + 2V\left[\frac{1}{4}\right] = \frac{\pi^2\hbar^2}{2m} + \frac{V}{2} = \frac{h^2}{8m} + \frac{V}{2}$$

$$H_{22} = 2\int_0^1 dx\,(\sin 2\pi x)\hat{H}(\sin 2\pi x)$$

$$= \frac{4\pi^2\hbar^2}{m}\int_0^1 dx \sin^2 2\pi x + 2V\int_0^1 dx\ x\sin^2 2\pi x$$

$$= \frac{\hbar^2}{2m} + 2V\left[\frac{1}{4}\right] = \frac{h^2}{2m} + \frac{V}{2}$$

$$H_{12} = H_{21} = 2\int_0^1 dx\,(\sin 2\pi x)\hat{H}(\sin\pi x)$$

$$= \frac{\pi^2\hbar^2}{m}\int_0^1 dx \sin 2\pi x \sin\pi x + 2V\int_0^1 dx\ x\sin 2\pi x\sin\pi x$$

The first integral here vanishes by orthogonality. To evaluate the second integral, we use the trigonometric identity

$$\sin\pi x\sin 2\pi x = (1/2)\cos\pi x - (1/2)\cos 3\pi x$$

and write

$$H_{12} = V\int_0^1 dx\ x[\cos\pi x - \cos 3\pi x]$$

$$= V\left|\frac{\cos\pi x}{\pi^2} + \frac{x\sin\pi x}{\pi} - \frac{\cos 3\pi x}{9\pi^2} - \frac{x\sin 3\pi x}{3\pi}\right|_0^1$$

$$= V\left[-\frac{2}{\pi^2} + \frac{2}{9\pi^2}\right] = -\frac{16V}{9\pi^2}$$

The secular determinantal equation is

$$\begin{vmatrix} \frac{h^2}{8m} + \frac{V}{2} - E & -\frac{16V}{9\pi^2} \\ -\frac{16V}{9\pi^2} & \frac{h^2}{2m} + \frac{V}{2} - E \end{vmatrix} = 0$$

The two roots to this equation are

$$E = \frac{1}{2}\left[\frac{5h^2}{8m} + V \pm \left[\frac{9h^4}{64m^2} + \frac{1024V^2}{81\pi^2}\right]^{1/2}\right]$$

In order to compare this result to the result obtained by perturbation theory in Section 7-2, we must linearize this result in V. Taking the smaller root, we have

$$E = \frac{1}{2}\left[\frac{5h^2}{8m} + V - \frac{3h^2}{8m}\left[1 + \frac{64m^2}{9h^4}\times\frac{1024V^2}{81\pi^2}\right]^{1/2}\right]$$

$$= \frac{1}{2}\left[\frac{5h^2}{8m} + V - \frac{3h^2}{8m}(1 + O(V^2))\right]$$

$$= \frac{h^2}{8m} + \frac{V}{2} + O(V^2)$$

which is Eq. 7-26. To obtain this result, we used the expansion

$$(1 + x)^{1/2} = 1 + (x/2) + O(x^2)$$

7-20. We must first evaluate the $H_{ij}$

$$H_{11} = \frac{h^2}{8ma^2} + V_1 a\left[\frac{1}{4} + \frac{1}{\pi^2}\right] \qquad \text{(Example 7-4)}$$

$$H_{22} = \frac{2}{a}\int_0^a \sin\frac{3\pi x}{a}\left[-\frac{\hbar^2}{2m}\frac{d^2}{dx^2}\right]\sin\frac{3\pi x}{a}dx$$

$$+ \frac{2V_1}{a}\int_0^{a/2} dx\ x\ \sin^2\frac{3\pi x}{a}$$

$$+ \frac{2V_1}{a}\int_{a/2}^{a} dx\,(a - x)\sin^2\frac{3\pi x}{a}$$

$$= \frac{9h^2}{8ma^2} + \frac{2V_1}{a}\left|\frac{x^2}{4} - \frac{x\sin\frac{6\pi x}{a}}{12\pi/a} - \frac{\cos\frac{6\pi x}{a}}{72\pi^2/a^2}\right|_0^{a/2}$$

$$+ 2V_1\left|\frac{x}{2} - \frac{\sin\frac{6\pi x}{a}}{12\pi/a}\right|_{a/2}^{a}$$

$$- \frac{2V_1}{a}\left|\frac{x^2}{4} - \frac{x\sin\frac{6\pi x}{a}}{12\pi/a} - \frac{\cos\frac{6\pi x}{a}}{72\pi^2/a^2}\right|_{a/2}^{a}$$

$$= \frac{9h^2}{8ma^2} + \frac{2V_1}{a}\left[\frac{a^2}{16} + \frac{a^2}{36\pi^2}\right] + 2V_1\left[\frac{a}{4}\right]$$

$$- \frac{2V_1}{a}\left[\frac{3a^2}{16} - \frac{a^2}{36\pi^2}\right]$$

$$= \frac{9h^2}{8ma^2} + V_1 a\left[\frac{1}{4} + \frac{1}{9\pi^2}\right]$$

$$H_{12} = H_{21} = \frac{2}{a}\int_0^a \sin\frac{\pi x}{a}\left[- \frac{\hbar^2}{2m}\frac{d^2}{dx^2}\right]\sin\frac{3\pi x}{a}dx$$

$$+ \frac{2V_1}{a}\int_0^{a/2} dx\ x\ \sin\frac{\pi x}{a}\sin\frac{3\pi x}{a}$$

$$+ \frac{2V_1}{a}\int_{a/2}^{a} dx\,(a - x)\sin\frac{\pi x}{a}\sin\frac{3\pi x}{a}$$

The first integral vanishes due to the orthogonality of $\sin(\pi x/a)$ and $\sin(3\pi x/a)$. In evaluating the second two integrals, it is convenient to use the trigonometric identity

$$\sin\frac{\pi x}{a}\sin\frac{3\pi x}{a} = \frac{1}{2}\cos\frac{2\pi x}{a} - \frac{1}{2}\cos\frac{4\pi x}{a}$$

Using this identity, we get

$$H_{12} = H_{21}$$

$$= \frac{V_1}{a}\left|\frac{\cos\frac{2\pi x}{a}}{4\pi^2/a^2} + \frac{x\sin\frac{2\pi x}{a}}{2\pi/a} - \frac{\cos\frac{4\pi x}{a}}{16\pi^2/a^2} - \frac{x\sin\frac{4\pi x}{a}}{4\pi/a}\right|_0^{a/2}$$

$$+ V_1 \left| \frac{\sin\frac{2\pi x}{a}}{2\pi/a} - \frac{\sin\frac{4\pi x}{a}}{4\pi/a} \right|_{a/2}^{a}$$

$$- \frac{V_1}{a} \left| \frac{\cos\frac{2\pi x}{a}}{4\pi^2/a^2} + \frac{x\sin\frac{2\pi x}{a}}{2\pi/a} - \frac{\cos\frac{4\pi x}{a}}{16\pi^2/a^2} - \frac{x\sin\frac{4\pi x}{a}}{4\pi/a} \right|_{a/2}^{a}$$

$$= \frac{V_1}{a}\left[- \frac{a^2}{2\pi^2}\right] + V_1(0) - \frac{V_1}{a}\left[\frac{a^2}{2\pi^2}\right]$$

$$= - \frac{V_1 a}{\pi^2}$$

The secular determinantal equation is

$$\begin{vmatrix} \frac{h^2}{8ma^2} + V_1 a\left[\frac{1}{4} + \frac{1}{\pi^2}\right] - E & - \frac{V_1 a}{\pi^2} \\ - \frac{V_1 a}{\pi^2} & \frac{9h^2}{8ma^2} + V_1 a\left[\frac{1}{4} + \frac{1}{9\pi^2}\right] - E \end{vmatrix} = 0$$

The two roots to the secular equation turn out to be (to $O(V^2)$, see Problem 7-19)

$$E = \frac{h^2}{8ma^2} + V_1 a\left[\frac{1}{4} + \frac{1}{\pi^2}\right]$$

$$E = \frac{9h^2}{8ma^2} + V_1 a\left[\frac{1}{4} + \frac{1}{9\pi^2}\right]$$

**7-21.** For a harmonic oscillator,

$$\psi_0(x) = \left[\frac{\alpha}{\pi}\right]^{1/4} e^{-\alpha x^2/2}$$

and

$$\psi_2(x) = \left[\frac{1}{8}\right]^{1/2}\left[\frac{\alpha}{\pi}\right]^{1/4}(4\alpha x^2 - 2)e^{-\alpha x^2/2}$$

We must first calculate the $H_{ij}$.

$$H_{11} = \int_{-\infty}^{\infty} \psi_0(x)\hat{H}\psi_0(x)dx$$

$$= \int_{-\infty}^{\infty} \psi_0(x)\left[- \frac{\hbar^2}{2\mu}\frac{d^2}{dx^2} + \frac{k}{2}x^2 + \frac{\gamma}{6}x^3 + \frac{b}{24}x^4\right]\psi_0(x)dx$$

$$= E_0 + \frac{\gamma}{6}\langle x^3\rangle_{00} + \frac{\delta}{24}\langle x^4\rangle_{00}$$

where $E_0 = (1/2)h\nu$.

The quantity $\langle x^3\rangle_{00}$ vanishes by symmetry (the integrand is an odd function), and

$$\langle x^4\rangle_{00} = \left[\frac{\alpha}{\pi}\right]^{1/2} 2\int_0^\infty dx\ x^4 e^{-\alpha x^2}$$

$$= \frac{3}{4\alpha^2}$$

$$H_{22} = \int_{-\infty}^{\infty}\psi_2(x)\left[-\frac{\hbar^2}{2\mu}\frac{d^2}{dx^2} + \frac{k}{2}x^2 + \frac{\gamma}{6}x^3 + \frac{\delta}{24}x^4\right]\psi_2(x)\,dx$$

$$= E_2 + \frac{\gamma}{6}\langle x^3\rangle_{22} + \frac{\delta}{24}\langle x^4\rangle_{22}$$

where $E_2 = (5/2)h\nu$. The quantity $\langle x^3\rangle_{22}$ vanishes by symmetry (the integrand is an odd function) and

$$\langle x^4\rangle_{22} = \frac{1}{8}\left[\frac{\alpha}{\pi}\right]^{1/2} 2\int_0^\infty dx\ x^4(4\alpha x^2 - 2)^2 e^{-\alpha x^2}$$

$$= \frac{1}{4}\left[\frac{\alpha}{\pi}\right]^{1/2}\int_0^\infty dx\,(16\alpha^2 x^8 - 16\alpha x^6 + 4x^4)e^{-\alpha x^2}$$

$$= \frac{1}{4}\left[\frac{\alpha}{\pi}\right]^{1/2}\left[\frac{105}{2\alpha^2} - \frac{15}{\alpha^2} + \frac{3}{2\alpha^2}\right]\left[\frac{\pi}{\alpha}\right]^{1/2}$$

$$= \frac{39}{4\alpha^2}$$

$$H_{12} = H_{21}$$

$$= \int_{-\infty}^{\infty}\psi_2(x)\left[-\frac{\hbar^2}{2\mu}\frac{d^2}{dx^2} + \frac{k}{2}x^2 + \frac{\gamma}{6}x^3 + \frac{\delta}{24}x^4\right]\psi_0(x)\,dx$$

$$= 0 + \frac{\gamma}{6}\langle x^3\rangle_{02} + \frac{\delta}{24}\langle x^4\rangle_{02}$$

The zero occurs because of the orthogonality of $\psi_0(x)$ and $\psi_2(x)$. The quantity $\langle x^3\rangle_{02}$ vanishes by symmetry (the integrand is an odd function) and

$$\langle x^4\rangle_{02} = \frac{1}{8^{1/2}}\left[\frac{\alpha}{\pi}\right]^{1/2} 2\int_0^\infty dx\ x^4(4\alpha x^2 - 2)e^{-\alpha x^2}$$

$$= \left[\frac{\alpha}{2\pi}\right]^{1/2} \left[\frac{15}{4\alpha^2} - \frac{3}{4\alpha^2}\right]\left[\frac{\pi}{\alpha}\right]^{1/2}$$

$$= \frac{3}{\sqrt{2}\alpha}$$

Thus we have

$$H_{11} = E_0 + \frac{\delta}{24}\left[\frac{3}{4\alpha^2}\right] = E_0 + \frac{\delta}{32\alpha^2}$$

$$H_{22} = E_2 + \frac{\delta}{24}\left[\frac{39}{4\alpha^2}\right] = E_2 + \frac{13\delta}{32\alpha^2}$$

$$H_{12} = H_{21} = \frac{\delta}{24}\left[\frac{3}{\sqrt{2}\alpha^2}\right] = \frac{\delta}{8\sqrt{2}\alpha^2}$$

The determinantal secular equation is

$$\begin{vmatrix} E_0 + \dfrac{\delta}{32\alpha^2} - E & \dfrac{\delta}{8\sqrt{2}\alpha^2} \\ \dfrac{\delta}{8\sqrt{2}\alpha^2} & E_2 + \dfrac{13\delta}{32\alpha^2} - E \end{vmatrix} = 0$$

The roots to this equation are

$$E = \frac{1}{2}\left[E_0 + E_2 + \frac{7\delta}{16\alpha^2}\right]$$
$$\pm \frac{1}{2}\left[\left[E_2 - E_0 + \frac{3\delta}{8\alpha^2}\right]^2 + \frac{\delta^2}{32\alpha^4}\right]^{1/2}$$

For small values of $\delta$, we may write this as

$$E = \frac{1}{2}\left[E_0 + E_2 + \frac{7\delta}{16\alpha^2}\right] \pm \frac{1}{2}\left[E_2 - E_0 + \frac{3\delta}{8\alpha^2}\right] + O(\delta^2)$$

The negative sign will give the lower root

$$E = E_0 + \frac{\delta}{32\alpha^2} + O(\delta^2)$$

This is the same result that is obtained by first-order perturbation theory.

**7-22.** We follow the procedure on pages 272 and 273. We calculate E for the given trial function and then expand E as a power series in $\varepsilon^2$. According to Eq. 7-64, the coefficient of $\varepsilon^2$ is $-\alpha/2$. The Hamiltonian is

$$\hat{H} = \hat{H}^{(o)} + e\mathcal{E}r\cos\theta$$

where $\hat{H}^{(o)}$ is the Hamiltonian for a hydrogen atom in the absence of any fields. Let's evaluate the $H_{ij}$.

$$H_{11} = \int d\tau\psi_{1s}\hat{H}\psi_{1s}$$

$$= \int d\tau\psi_{1s}[\hat{H}^{(o)} + e\mathcal{E}r\cos\theta]\psi_{1s}$$

We use the fact that

$$\hat{H}^{(o)}\psi_{1s} = E_1\psi_{1s}$$

to get

$$H_{11} = E_1 + e\mathcal{E}\int d\tau\psi_{1s}r\cos\theta\psi_{1s}$$

where

$$E_1 = -\frac{e^2}{8\pi\varepsilon_0 a_0} = -\frac{e^2}{2\kappa_0 a_0}$$

The second term in $H_{11}$ vanishes because of the integration over $\theta$

$$\int_0^\pi d\theta\sin\theta\cos\theta = \int_{-1}^{1} dx\ x = 0$$

and so

$$H_{11} = -\frac{e^2}{2\kappa_0 a_0}$$

The evaluation of $H_{22}$ is similar.

$$H_{22} = \int d\tau\psi_{3p_z}[\hat{H}^{(o)} + e\mathcal{E}r\cos\theta]\psi_{3p_z}$$

$$= E_3 + e\mathcal{E}\int d\tau\psi_{3p_z}r\cos\theta\psi_{3p_z}$$

where

$$E_3 = -\frac{e^2}{72\pi\varepsilon_0 a_0} = -\frac{e^2}{18\kappa_0 a_0}$$

The second term vanishes because of the integration over $\theta$

$$\int_0^\pi d\theta\sin\theta\cos^3\theta = \int_{-1}^{1} dx\ x^3 = 0$$

Therefore

$$H_{22} = -\frac{e^2}{18\kappa_0 a_0}$$

We now evaluate $H_{12} = H_{21}$

$$H_{12} = \int d\tau \psi_{1s}[\hat{H}^{(0)} + e\mathcal{E}r\cos\theta]\psi_{3p_z}$$

$$= 0 + e\mathcal{E}\int d\tau \psi_{1s} r\cos\theta \psi_{3p_z}$$

where the zero is due to the orthogonality of $\psi_{1s}$ and $\psi_{3p_z}$. Using the information in Table 6-5, we have ($\sigma = r/a_0$).

$$H_{12} = \frac{\sqrt{2}e\mathcal{E}}{81\pi a_0^3}\int_0^\infty dr\ r^3(6\sigma - \sigma^2)e^{-4\sigma/3} \int_0^{2\pi} d\phi \int_0^\pi d\theta \sin\theta\cos^2\theta$$

$$= \frac{4\sqrt{2}e\mathcal{E}}{3^5 a_0^3}\int_0^\infty dr\ r^3(6\sigma - \sigma^2)e^{-4\sigma/3}$$

$$= \frac{4\sqrt{2}e\mathcal{E}a_0}{3^5}\int_0^\infty d\sigma\ \sigma^3(6\sigma - \sigma^2)e^{-4\sigma/3}$$

$$= \frac{4\sqrt{2}e\mathcal{E}a_0}{3^5}\left[6\cdot 24\cdot \left[\frac{3}{4}\right]^5 - 120\left[\frac{3}{4}\right]^6\right]$$

$$= \frac{1}{\sqrt{2}}\left[\frac{3}{4}\right]^3 e\mathcal{E}a_0$$

The secular determinantal equation is

$$\begin{vmatrix} -\dfrac{e^2}{2\kappa_0 a_0} - E & \dfrac{1}{\sqrt{2}}\left[\dfrac{3}{4}\right]^3 e\mathcal{E}a_0 \\ \dfrac{1}{\sqrt{2}}\left[\dfrac{3}{4}\right]^3 e\mathcal{E}a_0 & -\dfrac{e^2}{18\kappa_0 a_0} - E \end{vmatrix} = 0$$

The two roots to this equation are

$$E = -\frac{5e^2}{18\kappa_0 a_0} \pm \frac{1}{2}\left[\frac{16e^4}{81\kappa_0^2 a_0^2} + 2\left[\frac{3}{4}\right]^6 e^2\mathcal{E}^2 a_0^2\right]^{1/2}$$

$$= -\frac{5e^2}{18\kappa_0 a_0} \pm \frac{4e^2}{18\kappa_0 a_0}\left[1 + \frac{3^{10}}{2^{15}}\frac{\kappa_0^2 a_0^4 \mathcal{E}^2}{e^2}\right]^{1/2}$$

We now expand the square root term according to

$$(1 + x)^{1/2} = 1 + \frac{1}{2}x + \cdots$$

and take the negative sign to get

$$E = -\frac{5e^2}{18\kappa_0 a_0} - \frac{4e^2}{18\kappa_0 a_0} - \frac{3^8}{2^{15}}\kappa_0 a_0^3 \mathcal{E}^2 + \cdots$$

$$= -\frac{e^2}{2\kappa_0 a_0} - 0.200\ \kappa_0 a_0^3 \mathcal{E}^2 + \cdots$$

The coefficient of $\mathcal{E}^2$ is $-\alpha/2$, and so we find that

$$\alpha = 0.400\ \kappa_0 a_0^3$$

This result is dicussed at the end of Section 7-5.

**7-23.** We must evaluate the elements $H_{ij}$.

$$H_{11} = \int \psi_0^{(0)*} \hat{H} \psi_0^{(0)} d\tau = \int \psi_0^{(0)*} [\hat{H}^{(0)} + \hat{H}^{(1)}] \psi_0^{(0)} d\tau$$

$$= E_0^{(0)} + E_0^{(1)}$$

$$H_{22} = \int \psi_1^{(0)*} [\hat{H}^{(0)} + \hat{H}^{(1)}] \psi_1^{(0)} d\tau$$

$$= E_1^{(0)} + E_1^{(1)}$$

$$H_{12} = H_{21} = \int \psi_1^{(0)*} [\hat{H}^{(0)} + \hat{H}^{(1)}] \psi_0^{(0)} d\tau$$

$$= 0 + H_{12}$$

Therefore, the secular determinantal equation is

$$\begin{vmatrix} E_0^{(0)} + E_0^{(1)} - E & H_{12} \\ H_{12} & E_1^{(0)} + E_1^{(1)} - E \end{vmatrix} = 0$$

The two roots to this quadratic equation are

$$E = \frac{E_0^{(0)} + E_0^{(1)} + E_1^{(0)} + E_1^{(1)}}{2}$$

$$\pm \frac{1}{2}\left[ \left[E_0^{(0)} + E_0^{(1)} - E_1^{(0)} - E_1^{(1)}\right]^2 + 4H_{12}^2 \right]$$

$$= \frac{E_0^{(0)} + E_0^{(1)} + E_1^{(0)} + E_1^{(1)}}{2}$$

$$\pm \frac{1}{2}\left[E_0^{(0)} + E_0^{(1)} - E_1^{(0)} - E_1^{(1)}\right]$$

$$\times \left[1 + \frac{4H_{12}^2}{(E_0^{(0)} + E_0^{(1)} - E_1^{(0)} - E_1^{(1)})^2}\right]^{1/2}$$

Assuming that $H_{12}$ is small, we expand the term in brackets using the formula

$$(1 + x)^{1/2} = 1 + \frac{x}{2} + \cdots$$

to get

$$E = \frac{E_0^{(0)} + E_0^{(1)} + E_1^{(0)} + E_1^{(1)}}{2}$$

$$\pm \frac{(E_0^{(0)} + E_0^{(1)} - E_1^{(0)} - E_1^{(1)})}{2}$$

$$\pm \frac{H_{12}^2}{(E_0^{(0)} + E_0^{(1)} - E_1^{(0)} - E_1^{(1)})} + \cdots$$

Because the 0 state has a lower energy than the 1 state, we choose the positive signs

$$E = E_0^{(0)} + E_0^{(1)}$$

$$+ \frac{H_{12}^2}{(E_0^{(0)} + E_0^{(1)} - E_1^{(0)} - E_1^{(1)})} + \cdots$$

If we had chosen the negative signs, then the energy would have been $E_1^{(0)}$ to lowest order.
As stated in the problem we cannot identify this third term as a second-order term because although the numerator is second-order, the denominator is both zero-order and first-order. We use the expansion

$$\frac{1}{E_0^{(0)} + E_0^{(1)} - E_1^{(0)} - E_1^{(1)}}$$

$$= \frac{1}{E_0^{(0)} - E_1^{(0)}} \left[\frac{1}{1 + \dfrac{E_0^{(1)} - E_1^{(1)}}{E_0^{(0)} - E_1^{(0)}}}\right]$$

$$= \frac{1}{E_0^{(0)} - E_1^{(0)}}\left[1 - \frac{E_0^{(1)} - E_1^{(1)}}{E_0^{(0)} - E_1^{(0)}} + \cdots\right]$$

to finally obtain

$$E = E_0^{(0)} - E_0^{(1)} + \frac{H_{12}^2}{E_0^{(0)} - E_1^{(0)}} + \cdots$$

or

$$E_0^{(2)} = \frac{H_{12}^2}{E_0^{(0)} - E_1^{(0)}}$$

If we had used more terms in the trial function, then we would arrive at Eq. 15 of Problem 7-24.

**7-24.** If we substitute Eqs. 3, 4 and 5 of Problem 7-24 into Eq. 1, then we obtain

$$[\hat{H}^{(0)} + \lambda\hat{H}^{(1)}][\psi_n^{(0)} + \lambda\psi_n^{(1)} + \lambda^2\psi_n^{(2)} + \cdots]$$
$$= E_n^{(0)} + \lambda E_n^{(1)} + \lambda^2 E_n^{(2)} + \cdots]$$
$$[\psi_n^{(0)} + \lambda\psi_n^{(1)} + \lambda^2\psi_n^{(2)} + \cdots]$$

We carry out the multiplication and collect like powers of $\lambda$ on each side of the equation to obtain

$$\hat{H}^{(0)}\psi_n^{(0)} + \lambda[\hat{H}^{(0)}\psi_n^{(1)} + \hat{H}^{(1)}\psi_n^{(0)}] + \lambda^2[\hat{H}^{(0)}\psi_n^{(2)} + \hat{H}^{(1)}\psi_n^{(1)}] + \cdots$$
$$= E_n^{(0)}\psi_n^{(0)} + \lambda[E_n^{(1)}\psi_n^{(0)} + E_n^{(0)}\psi_n^{(1)}] + \lambda^2[E_n^{(0)}\psi_n^{(2)} + E_n^{(1)}\psi_n^{(1)} + E_n^{(2)}\psi_n^{(0)}] + \cdots$$

Equating the coefficients of each power of $\lambda$ gives

$$\hat{H}^{(0)}\psi_n^{(0)} = E_n^{(0)}\psi_n^{(0)} \qquad \text{(zero-order)}$$

$$\hat{H}^{(0)}\psi_n^{(1)} + \hat{H}^{(1)}\psi_n^{(0)} = E_n^{(1)}\psi_n^{(0)} + E_n^{(0)}\psi_n^{(1)} \qquad \text{(first-order)}$$

$$\hat{H}^{(0)}\psi_n^{(2)} + \hat{H}^{(1)}\psi_n^{(1)} = E_n^{(2)}\psi_n^{(0)} + E_n^{(1)}\psi_n^{(1)} + E_n^{(0)}\psi_n^{(2)} \qquad \text{(second-order)}$$

Multiply the first-order equation from the left by $\psi_n^{(0)*}$ and integrate to get

$$\int\psi_n^{(0)*}\hat{H}^{(0)}\psi_n^{(1)}d\tau + \int\psi_n^{(0)*}\hat{H}^{(1)}\psi_n^{(0)}d\tau$$
$$= \int\psi_n^{(0)*}E_n^{(1)}\psi_n^{(0)}d\tau + \int\psi_n^{(0)*}E_n^{(0)}\psi_n^{(1)}d\tau$$

Using the Hermitian property of $\hat{H}^{(0)}$, we see that

$$\int\psi_n^{(0)*}\hat{H}^{(0)}\psi_n^{(1)}d\tau = \int(\hat{H}^{(0)}\psi_n^{(0)})^*\psi_n^{(1)}d\tau$$
$$= E_n^{(0)}\int\psi_n^{(0)*}\psi_n^{(1)}d\tau$$

which cancels the second term on the right-hand side.

Thus we are left with

$$E_n^{(1)} = \int \psi_n^{(0)*}\hat{H}^{(1)}\psi_n^{(0)}d\tau$$

We now write

$$\psi_n^{(1)} = \sum_j a_{nj}\psi_j^{(0)}$$

and substitute into the first-order equation to get

$$(\hat{H}^{(0)} - E_n^{(0)})\psi_n^{(1)} = \sum_j a_{nj}[\hat{H}^{(0)} - E_n^{(0)}]\psi_j^{(0)}$$
$$= E_n^{(1)}\psi_n^{(0)} - \hat{H}^{(1)}\psi_n^{(0)}$$

Multiply from the left by $\psi_k^{(0)*}$ and integrate to obtain

$$\sum_j a_{nj}\int \psi_k^{(0)*}\left[\hat{H}^{(0)} - E_n^{(0)}\right]\psi_j^{(0)}d\tau$$
$$= E_n^{(1)}\int \psi_k^{(0)*}\psi_n^{(0)}d\tau - \int \psi_k^{(0)*}\hat{H}^{(1)}\psi_n^{(0)}d\tau$$

First consider the case in which k = n. Then

$$\sum_j a_{nj}\int \psi_n^{(0)*}[\hat{H}^{(0)} - E_n^{(0)}]\ \psi_j^{(0)}d\tau$$
$$= \sum_j a_{nj}\int \left[[\hat{H}^{(0)} - E_n^{(0)}]\psi_n^{(0)}\right]^*\psi_j d\tau = 0$$

and

$$E_n^{(1)} = \int \psi_n^{(0)*}\hat{H}^{(1)}\psi_n^{(0)}d\tau$$

Now consider the case in which $k \neq n$. Then

$$\sum_j a_{nj}\int \psi_k^{(0)*}\left[\hat{H}^{(0)} - E_n^{(0)}\right]\psi_j^{(0)}d\tau$$
$$= \sum_j a_{nj}\int \psi_k^{(0)*}\left[E_j^{(0)} - E_n^{(0)}\right]\psi_j^{(0)}d\tau$$
$$= \sum_j a_{nj}[E_j^{(0)} - E_n^{(0)}]\int \psi_k^{(0)*}\ \psi_j^{(0)}d\tau$$
$$= \sum_j a_{nj}(E_j^{(0)} - E_n^{(0)})\delta_{kj}$$
$$= a_{nk}(E_k^{(0)} - E_n^{(0)}) \qquad k \neq n$$
$$E_n^{(1)}\int \psi_k^{(0)*}\psi_n^{(0)}d\tau = E_n^{(1)}\delta_{kn} = 0 \qquad k \neq n$$

and so

$$a_{nk} = \frac{\int \psi_k^{(0)*} \hat{H}^{(1)} \psi_n^{(0)} d\tau}{E_n^{(0)} - E_k^{(0)}} \qquad k \neq n$$

$$= \frac{H_{kn}^{(1)}}{E_n^{(0)} - E_k^{(0)}} \qquad k \neq n$$

The normalization condition of $\psi_n$ in Eq. 4 is

$$\int \psi_n^* \psi_n d\tau = \int [\psi_n^{(0)} + \lambda\psi_n^{(1)} + \cdots]^* \times [\psi_n^{(0)} + \lambda\psi_n^{(1)} + \cdots] d\tau$$

$$= \int \psi_n^{(0)*} \psi_n^{(0)} d\tau + \lambda \int \psi_n^{(1)*} \psi_n^{(0)} d\tau + \lambda \int \psi_n^{(0)*} \psi_n^{(1)} d\tau + O(\lambda^2)$$

$$= 1 + \lambda \int \psi_n^{(1)*} \psi_n^{(0)} d\tau + \lambda \int \psi_n^{(0)*} \psi_n^{(1)} d\tau + O(\lambda^2)$$

For $\psi_n$ to be normalized through terms linear in $\lambda$, we must have that

$$\int \psi_n^{(1)*} \psi_n^{(0)} d\tau = \int \psi_n^{(0)*} \psi_n^{(1)} d\tau = 0$$

or that $\psi_n^{(1)}$ be orthogonal to $\psi_n^{(0)}$. Substituting Eq. 9 into this orthogonality condition gives

$$\sum_j a_{nj} \int \psi_j^{(0)*} \psi_n^{(0)} d\tau = \sum_j a_{nj} \delta_{nj} = a_{nn} = 0$$

Thus, the wave function to first order is

$$\psi_n = \psi_n^{(0)} + \lambda \sum_{j \neq n} \frac{H_{jn}^{(1)}}{E_n^{(0)} - E_j^{(0)}} \psi_j^{(0)} \qquad \text{(first-order)}$$

To find an expression for $E_n^{(2)}$, we start with

$$\hat{H}^{(0)} \psi_n^{(2)} + \hat{H}^{(1)} \psi_n^{(1)} = E_n^{(2)} \psi_n^{(0)} + E_n^{(1)} \psi_n^{(1)} + E_n^{(0)} \psi_n^{(2)}$$

We substitute

$$\psi_n^{(2)} = \sum_s b_{ns} \psi_s^{(0)}$$

into the above second-order equation to get

$$\sum_s b_{ns}E_s^{(0)}\psi_s^{(0)} + \hat{H}^{(1)}\sum_j a_{nj}\psi_j^{(0)}$$

$$= E_n^{(2)}\psi_n^{(0)} + E_n^{(1)}\sum_j a_{nj}\psi_n^{(0)} + E_n^{(0)}\sum_s b_{ns}\psi_s^{(0)}$$

Multiply from the left by $\psi_m^{(0)*}$ and integrate to get

$$b_{nm}E_m^{(0)} + \sum_j a_{nj}H_{mj}^{(1)} = E_n^{(2)}\delta_{mn} + E_n^{(1)}a_{mn} + E_n^{(0)}b_{mn}$$

If n = m, then

$$E_n^{(2)} = \sum_j a_{nj}H_{nj}^{(1)} = \sum_j \frac{H_{nj}^{(1)}H_{jn}^{(1)}}{E_n^{(0)} - E_j^{(0)}} \qquad n \neq j$$

If n ≠ m, then the above expression becomes an expression for $b_{nj}$.

# CHAPTER EIGHT

## ATOMS

**8-1.** The ground-state energy of a helium ion is

$$E_1 = -\frac{me^4Z^2}{32\pi^2\varepsilon_0^2\hbar^2}$$

If we let $m = 1$, $e = 1$, $4\pi\varepsilon_0 = 1$, and $\hbar = 1$, then we obtain

$$E_1 = -(Z^2/2) = -2 \text{ au} \qquad (Z = 2)$$

**8-2.** In atomic units, $e = 1$, $4\pi\varepsilon_0 = 1$ and the Bohr radius, $a_0$, equals unity. Thus

$$V = \frac{e}{4\pi\varepsilon_0 a_0} = 1 \text{ au}$$

**8-3.** Start with Eq. 1-23 with $n = 1$ and $r = a_0$,

$$v = \frac{\hbar}{ma_0} = \frac{\hbar}{m}\frac{me^2}{\kappa_0\hbar^2} = \frac{e^2}{\kappa_0\hbar}$$

$$= \frac{(1.6022 \times 10^{-19} \text{ C})^2}{(1.1126 \times 10^{-10} \text{ C}^2\cdot\text{s}^2\cdot\text{kg}^{-1}\cdot\text{m}^{-3}) \times (1.0546 \times 10^{-34} \text{ J}\cdot\text{s})}$$

$$= 2.188 \times 10^6 \text{ m}\cdot\text{s}^{-1}$$

**8-4.** From Problem 8-3, the speed of an electron in the first Bohr orbit is

$$v = \frac{e^2}{\kappa_0\hbar}$$

The time for an electron to travel one radian is

$$T = \frac{a_0}{v} = \frac{\kappa_0\hbar^2}{me^2}\frac{\kappa_0\hbar}{e^2} = \frac{\kappa_0^2\hbar^3}{me^4}$$

$$= \frac{(1.1126 \times 10^{-10} \text{ C}^2\cdot\text{s}^2\cdot\text{kg}^{-1}\cdot\text{m}^{-3}) \times (1.0546 \times 10^{-34} \text{ J}\cdot\text{s})^3}{(9.1095 \times 10^{-31} \text{ kg})(1.6022 \times 10^{-19} \text{ C})^4}$$

$$= 2.418 \times 10^{-17} \text{ s}$$

8-5.

$$c = \frac{(2.9979 \times 10^8 \text{ m·s}^{-1})}{(2.18764 \times 10^6 \text{ m·s}^{-1}\text{·au}^{-1})}$$

$$= 137.04 \text{ au}$$

8-6. We start with

$$E^{(1)} = \frac{Z^6}{\pi^2}\int_0^\infty \exp(-2Zr_1)4\pi r_1^2 dr_1 \int_0^\infty \exp(-2Zr_2) r_2^2 dr_2$$

$$\times \int_0^{2\pi} d\phi \int_0^\pi \frac{\sin\theta d\theta}{(r_1^2 + r_2^2 - 2r_1r_2\cos\theta)^{1/2}}$$

Let $x = \cos\theta$. Then $dx = -\sin\theta d\theta$ and the limits $(0,\pi)$ become $(+1, -1)$. Then

$$\int_0^\pi \frac{\sin\theta d\theta}{(r_1^2 + r_2^2 - 2r_1r_2\cos\theta)^{1/2}}$$

$$= \int_{-1}^{1} \frac{dx}{(r_1^2 + r_2^2 - 2r_1r_2x)^{1/2}}$$

$$= \left| {}_{-1}^{1} - \frac{(r_1^2 + r_2^2 - 2r_1r_2x)^{1/2}}{r_1r_2} \right|$$

When $x = 1$ we have $[(r_1 - r_2)^2]^{1/2} = \pm\,(r_1 - r_2)$ and when $x = -1$ we have $[(r_1 + r_2)^2]^{1/2} = \pm\,(r_1 + r_2)$. We must decide which signs to take. Let's set up a table to choose the value of

$$\frac{\pm\,(r_1 + r_2)}{r_1r_2} \pm \frac{(r_1 - r_2)}{r_1r_2}$$

| signs | | value |
|---|---|---|
| + | + | $2/r_2$ |
| + | − | $2/r_1$ |
| − | + | $-2/r_1$ |
| − | − | $-2/r_2$ |

We reject the two negative values because the integrand of $E^{(1)}$, and hence $E^{(1)}$ itself, is an intrinsically positive quantity.

Now let's decide how to choose the positive values. When $r_1$ is very large, $(r_1^2 + r_2^2 - 2r_1r_2x)^{1/2} \to r_1$, and

$$\int_{-1}^{1} \frac{dx}{(r_1^2 + r_2^2 - 2r_1r_2x)^{1/2}} \to \frac{2}{r_1}$$

When $r_2$ is very large, $(r_1^2 + r_2^2 - 2r_1r_2x)^{1/2} \to r_2$, and

$$\int_{-1}^{1} \frac{dx}{(r_1^2 + r_2^2 - 2r_1r_2x)^{1/2}} \to \frac{2}{r_2}$$

These are the two values that we found above for the integral, and so we must have that

$$\int_{-1}^{1} \frac{dx}{(r_1^2 + r_2^2 - 2r_1r_2x)^{1/2}} = \frac{2}{r_1} \qquad r_1 > r_2$$

$$= \frac{2}{r_2} \qquad r_1 < r_2$$

We now use this result to evaluate $E^{(1)}$.

$$E^{(1)} = 16Z^6\int_0^\infty \exp(-2Zr_1)r_1^2dr_1\left[\frac{1}{r_1}\int_0^{r_1}\exp(-2Zr_2)r_2^2dr_2 + \int_{r_1}^\infty \exp(-2Zr_2)r_2dr_2\right]$$

$$= 4Z^3\int_0^\infty \exp(-2Zr_1)r_1^2dr_1 \times \left[\frac{1}{r_1} - \exp(-2Zr_1)\left[Z + \frac{1}{r_1}\right]\right]$$

$$= 4Z^3\left[\frac{1}{4Z^2} - \frac{Z\cdot 2!}{(4Z)^3} - \frac{1}{16Z^2}\right]$$

$$= 4Z^3\left[\frac{5}{32Z^2}\right] = \frac{5Z}{8}$$

The energy through first order is

$$E = E^{(0)} + E^{(1)} = -Z^2 + \frac{5}{8}Z = -4 + \frac{5}{4} = -\frac{11}{4}\text{au}$$

**8-7.**

$$E(Z) = \iint d\mathbf{r}_1 d\mathbf{r}_2 \psi^* \hat{H}\psi$$

$$= \frac{Z^6}{\pi^2}\iint d\mathbf{r}_1 d\mathbf{r}_2 \exp(-Z(r_1+r_2))\left[-Z^2 + \frac{(Z-2)}{r_1} + \frac{(Z-2)}{r_2} + \frac{1}{r_{12}}\right]\exp(-Z(r_1+r_2))$$

where we have used the fact that

$$\left[-\frac{1}{2}\nabla_1{}^2 - \frac{Z}{r_1} - \frac{1}{2}\nabla_2{}^2 - \frac{Z}{r_2}\right]\exp(-Z(r_1 + r_2))$$

$$= \left[-\frac{Z^2}{2} - \frac{Z^2}{2}\right]\exp(-Z(r_1 + r_2))$$

The first integral in E(Z) yields $-Z^2$ since $\psi$ is normalized. The next two integrals are equal, and each one is given by

$$I = \frac{Z^3}{\pi}(Z - 2)4\pi\int_0^\infty dr\ re^{-2Zr}$$

$$= 4(Z - 2)Z^3\left[\frac{1}{4Z^2}\right] = Z(Z - 2)$$

The fourth integral is evaluated in Problem 8-6, and is equal to 5Z/8. Thus

$$E(Z) = -Z^2 + 2Z(Z - 2) + \frac{5}{8}Z = Z^2 - \frac{27}{8}Z$$

The condition $\frac{dE}{dZ} = 0$ gives $Z_{min} = \frac{27}{16}$

and

$$E_{min} = \left[\frac{27}{16}\right]^2 - \frac{27}{8}\left[\frac{27}{16}\right] = -\left[\frac{27}{16}\right]^2 = -2.8477 \text{ au}$$

The value of $Z_{min}$ is less than 2, the nuclear charge of helium, due to the screening of one electron from the nucleus by the other electron.

**8-8.** The radial part of a Slater orbital is $r^{n-1}e^{-\xi r}$. The normalization constant is given by

$$c^2\int_0^\infty r^{2n-2}e^{-2\xi r}r^2dr = c^2\frac{(2n)!}{(2\xi)^{2n+1}} = 1$$

or

$$c = \frac{(2\xi)^{n+(1/2)}}{\sqrt{(2n)!}}$$

**8-9.** The radial parts of Slater orbitals are everywhere positive and so an integral of their product can never be zero.

**8-10.** Start with Eq. 8-29

$$E = \iint d\mathbf{r}_1 d\mathbf{r}_2 \phi^*(\mathbf{r}_1)\phi^*(\mathbf{r}_2)\left[-\frac{1}{2}\nabla_1^2 - \frac{1}{2}\nabla_2^2 - \frac{Z}{r_1} - \frac{Z}{r_2} + \frac{1}{r_{12}}\right]\phi(\mathbf{r}_1)\phi(\mathbf{r}_2)$$

$$= \int d\mathbf{r}_1 \phi^*(\mathbf{r}_1)\left[-\frac{1}{2}\nabla_1^2 - \frac{Z}{r_1}\right]\phi(\mathbf{r}_1)\int d\mathbf{r}_2 \phi^*(\mathbf{r}_2)\phi(\mathbf{r}_2)$$

$$+ \int d\mathbf{r}_2 \phi^*(\mathbf{r}_2)\left[-\frac{1}{2}\nabla_2^2 - \frac{Z}{r_2}\right]\phi(\mathbf{r}_2)\int d\mathbf{r}_1 \phi^*(\mathbf{r}_1)\phi(\mathbf{r}_1)$$

$$+ \iint d\mathbf{r}_1 d\mathbf{r}_2 \phi^*(\mathbf{r}_1)\phi^*(\mathbf{r}_2)\frac{1}{r_{12}}\phi(\mathbf{r}_1)\phi(\mathbf{r}_2)$$

The integrals of $\phi^*(\mathbf{r}_i)\phi(\mathbf{r}_i)$ are unity by normalization, and so

$$E = I_1 + I_2 + J_{12}$$

**8-11.** Recall that the angular part of $\nabla^2$ is $\hat{L}^2$, and that the $Y_\ell^m(\theta,\phi)$ are eigenfunctions of $\hat{L}^2$ with eigenvalues $\hbar^2\ell(\ell+1)$, or simply $\ell(\ell+1)$ in atomic units.

**8-12.** Start with Eq. 8-36

$$\varepsilon_1 = \int d\mathbf{r}_1 \phi^*(\mathbf{r}_1)\hat{H}_1^{eff}(\mathbf{r}_1)\phi(\mathbf{r}_1)$$

$$= \int d\mathbf{r}_1 \phi^*(\mathbf{r}_1)\left[-\frac{1}{2}\nabla_1^2 - \frac{Z}{r_1} + U_1^{eff}(\mathbf{r}_1)\right]\phi(\mathbf{r}_1)$$

$$= I_1 + J_{12}$$

The sum of the orbital energies, $\varepsilon_1 + \varepsilon_2$, is

$$I_1 + I_2 + 2J_{12} \neq E = I_1 + I_2 + J_{12}$$

**8-13.** Refer to Figure 8-7.

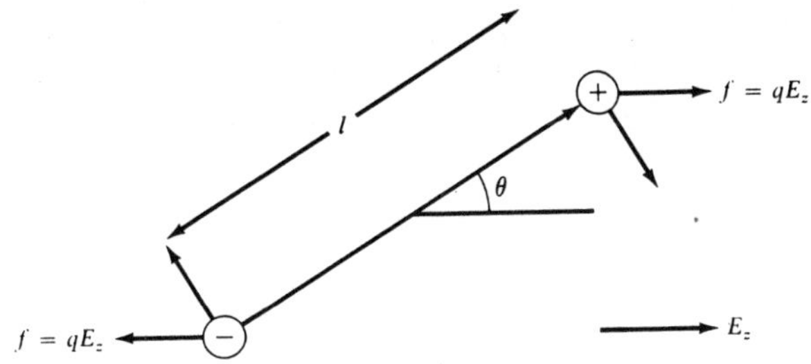


The force perpendicular to $\ell$ is given by

$$f_{\perp} = f\sin\theta = qE_z\sin\theta$$

The torque is given by

$$\tau = \ell f_{\perp} = \ell qE_z\sin\theta = \mu E_z\sin\theta$$

The energy of a dipole in an electric field $E_z$ is

$$\varepsilon = \int_0^\theta \mu E_z\sin\theta' d\theta' = -\mu E_z\cos\theta = -\mu_z E_z$$

**8-14.** The two states involved in the transition have spin angular momenta $\pm\hbar/2$, and so (compare to Eq. 6-141)

$$\Delta E = g(e\hbar/2m_p)B_z$$

$$= (5.5849)\left[\frac{(1.602\times10^{-19}\text{ C})\times(1.055\times10^{-34}\text{ J}\cdot\text{s})}{2(1.673\times10^{-27}\text{ kg})}\right](2.0\text{ T})$$

$$= 5.642\times10^{-26}\text{ J}$$

The frequency of the radiation for such an experiment is

$$\nu = \frac{\Delta E}{h} = 8.515\times10^{7}\text{ Hz} = 85.15\text{ MHz}$$

**8-15.**

$$\iint\Psi^*(1,2)\Psi(1,2)d\tau_1 d\tau_2$$

$$= \iint[1s\alpha^*(1)1s\beta^*(2) - 1s\alpha^*(2)1s\beta^*(1)] \times [1s\alpha(1)1s\beta(2) - 1s\alpha(2)1s\beta(1)]d\tau_1 d\tau_2$$

$$= \iint 1s\alpha^*(1)1s\beta^*(2)1s\alpha(1)1s\beta(2)d\tau_1 d\tau_2$$

$$- \iint 1s\alpha^*(1)1s\beta^*(2)1s\alpha(2)1s\beta(1)d\tau_1 d\tau_2$$

$$- \iint 1s\alpha^*(2)1s\beta^*(1)1s\alpha(1)1s\beta(2)d\tau_1 d\tau_2$$

$$+ \iint 1s\alpha^*(2)1s\beta^*(1)1s\alpha(2)1s\beta(1)d\tau_1 d\tau_2$$

$$= 1 - 0 - 0 + 1 = 2$$

where we have used Eqs. 8-46.

**8-16.** Suppose that the first two columns are the same. Then $a_{11} = a_{12}$, $a_{21} = a_{22}$, $a_{31} = a_{32}$, and

$$D = a_{11}\begin{vmatrix} a_{21} & a_{23} \\ a_{31} & a_{33} \end{vmatrix} - a_{11}\begin{vmatrix} a_{21} & a_{23} \\ a_{31} & a_{33} \end{vmatrix} + a_{13}\begin{vmatrix} a_{21} & a_{21} \\ a_{31} & a_{31} \end{vmatrix} = 0$$

Suppose that the second and third columns are the same. Then

$$D = a_{11}\begin{vmatrix} a_{22} & a_{22} \\ a_{32} & a_{32} \end{vmatrix} - a_{12}\begin{vmatrix} a_{21} & a_{22} \\ a_{31} & a_{32} \end{vmatrix} + a_{12}\begin{vmatrix} a_{21} & a_{22} \\ a_{31} & a_{32} \end{vmatrix} = 0$$

and so forth.

If we interchange columns 1 and 2, then $a_{11} \leftrightarrow a_{12}$, $\alpha_{21} \leftrightarrow a_{22}$, $a_{31} \leftrightarrow a_{32}$ and

$$D_{\text{interchange}} = a_{12}\begin{vmatrix} a_{21} & a_{23} \\ a_{31} & a_{33} \end{vmatrix} - a_{11}\begin{vmatrix} a_{22} & a_{23} \\ a_{32} & a_{33} \end{vmatrix} + a_{13}\begin{vmatrix} a_{22} & a_{21} \\ a_{32} & a_{31} \end{vmatrix}$$

$$= - D_{\text{original}}$$

The proof for the interchange of other columns or rows is similar.

**8-17.** This problem is similar to Problem 8-15.

**8-18.**

$$\psi = \frac{1}{\sqrt{2}}[1s(1)\alpha(1)1s(2)\beta(2) - 1s(1)\beta(1)1s(2)\alpha(2)]$$

$$= 1s(1)1s(2)\left[\frac{1}{\sqrt{2}}(\alpha(1)\beta(2) - \beta(1)\alpha(2)\right]$$

**8-19.** The expansion of an N × N determinant yields N! terms. The integral over each term will equal unity if the individual spin orbitals are normalized. See Problems 8-15 and 8-17 for the N = 2 case.

**8-20.**

$$\hat{S}_{z,total}\psi = \frac{1}{\sqrt{2}}(\hat{S}_{z1} + \hat{S}_{z2})[1s\alpha(1)1s\beta(2) - 1s\beta(1)1s\alpha(2)]$$

$$= \frac{1}{\sqrt{2}}\left[\left[\frac{\hbar}{2} - \frac{\hbar}{2}\right]1s\alpha(1)1s\beta(2) - \left[-\frac{\hbar}{2} + \frac{\hbar}{2}\right]1s\beta(1)1s\alpha(2)\right] = 0$$

Thus $\psi$ is an eigenfunction with eigenvalue zero. The expansion of the second Slater determinant in this problem involves six terms. The diagonal product is typical of each of these six terms. Consider then

$$(\hat{S}_{z1} + \hat{S}_{z2} + \hat{S}_{z3})1s\alpha(1)1s\beta(2)2s\alpha(3)$$

$$= \left[\frac{\hbar}{2} - \frac{\hbar}{2} + \frac{\hbar}{2}\right]1s\alpha(1)1s\beta(2)2s\alpha(3)$$

Thus $\psi$ is an eigenfuction with eigenvalue $+\hbar/2$, or 1/2 in atomic units.

**8-21.** Use the facts that

$$\hat{L}_z\psi_{n\ell m} = m\hbar\psi_{n\ell m}$$

and

$$\hat{S}_z\alpha = +\frac{\hbar}{2}\alpha$$

$$\hat{S}_z \beta = -\frac{\hbar}{2}\beta$$

Thus

$$\hat{L}_{z,\text{total}}\Psi = \frac{1}{\sqrt{2}}(\hat{L}_{z1} + \hat{L}_{z2})\left[\psi_{211}\alpha(1)\psi_{21-1}\beta(2) - \psi_{211}\alpha(2)\psi_{21-1}\beta(1)\right]$$

$$= \frac{1}{\sqrt{2}}\left[(+\hbar - \hbar)\psi_{211}\alpha(1)\psi_{21-1}\beta(2) - (-\hbar + \hbar)\psi_{211}\alpha(2)\psi_{21-1}\beta(1)\right]$$

$$= 0$$

Therefore $\Psi$ is an eigenfunction of $\hat{L}_{z,\text{total}}$ with eigenvalue 0.

Similarly,

$$\hat{S}_{z,\text{total}}\Psi = \frac{1}{\sqrt{2}}(\hat{S}_{z1} + \hat{S}_{z2})[\psi_{211}\alpha(1)\psi_{21-1}\beta(2) - \psi_{211}\alpha(2)\psi_{21-1}\beta(1)]$$

$$= \frac{1}{\sqrt{2}}\left[\left[+\frac{\hbar}{2} - \frac{\hbar}{2}\right]\psi_{211}\alpha(1)\psi_{21-1}\beta(2) - \left[-\frac{\hbar}{2} + \frac{\hbar}{2}\right]\psi_{211}\alpha(2)\psi_{21-1}\beta(1)\right]$$

$$= 0$$

Therefore $\Psi$ is also an eigenfunction of $\hat{S}_{z,\text{total}}$ with eigenvalue 0.

**8-22.** The three symmetric spin functions are

1. $\alpha(1)\alpha(2)$

4. $\beta(1)\beta(2)$

$\psi_+$ $\frac{1}{\sqrt{2}}[\alpha(1)\beta(2) + \alpha(2)\beta(1)]$

and the antisymmetric spin funciton is

$\psi_-$ $\frac{1}{\sqrt{2}}[\alpha(1)\beta(2) - \alpha(2)\beta(1)]$

The product of the spatial wave functions with the spin functions must be antisymmetric, and so we combine an antisymmetric spatial wave function with the symmetric spin functions and a symmetric spatial wave function with the single antisymmetric spin function. Thus, there are four allowable combinations

$$[\psi(1)\phi(2) + \psi(2)\phi(1)]\frac{1}{\sqrt{2}}[\alpha(1)\beta(2) - \alpha(2)\beta(1)]$$

$$[\psi(1)\phi(2) - \psi(2)\phi(1)][\alpha(1)\alpha(2)]$$

$$[\psi(1)\phi(2) - \psi(2)\phi(1)][\beta(1)\beta(2)]$$

$$[\psi(1)\phi(2) - \psi(2)\phi(1)]\frac{1}{\sqrt{2}}[\alpha(1)\beta(2) + \alpha(2)\beta(1)]$$

By operating on each of these wave functions with $\hat{S}_{z,total} = \hat{S}_{z1} + \hat{S}_{z2}$, we find that (see Problem 8-20) each wave function is an eigenfunction of $\hat{S}_{z,total}$, and that the eigenvalues are 0, $+\hbar$, $-\hbar$ and 0, respectively.

**8-23.** Problem 8-22 shows that there are four possible determinantal wave functions for a helium atom with a 1s2s electron configuration. We shall see that $\phi_1$, $\phi_2$, $\phi_3 - \phi_4$ and $\phi_3 + \phi_4$ correspond to the four antisymmetric wave functions given in Problem 8-22.

To derive the secular determinantal equation for this problem, we must determine the $H_{ij}$ and the $S_{ij}$. Let's look at each one in turn.

$$H_{11} = \iint d\tau_1 d\tau_2 \phi_1 \hat{H} \phi_1$$

But

$$\hat{H}\phi_1 = \left[-\frac{1}{2}\nabla_1^2 - \frac{2}{r_1} - \frac{1}{2}\nabla_2^2 - \frac{2}{r_2} + \frac{1}{r_{12}}\right] \frac{1}{\sqrt{2}}[1s\alpha(1)2s\alpha(2) - 1s\alpha(2)2s\alpha(1)]$$

$$= \left[-\frac{Z^2}{2} - \frac{Z^2}{8} + \frac{1}{r_{12}}\right]\frac{1}{\sqrt{2}}[1s\alpha(1)2s\alpha(2) - 1s\alpha(2)2s\alpha(1)]$$

where we have used the fact that the eigenvalue associated with a 1s orbital is $-Z^2/2$ and that associated with a 2s orbital is $-Z^2/8$ ($n = 2$).

For helium, $Z = 2$, and so

$$-\frac{Z^2}{2} - \frac{Z^2}{8} = -\frac{5}{2}\text{ au} = E_0$$

Thus

$$H_{11} = \frac{1}{2}\iint d\tau_1 d\tau_2 [1s\alpha^*(1)2s\alpha^*(2) - 1s\alpha^*(2)2s\alpha^*(1)] \times \left[E_0 + \frac{1}{r_{12}}\right][1s\alpha(1)2s\alpha(2) - 1s\alpha(2)2s\alpha(1)]$$

$$= \frac{1}{2}[E_0(1 - 0 - 0 + 1) + J - K - K + J]$$

$$= E_0 + J - K$$

The evaluation of $H_{22}$ is very similar to the evaluation of $H_{11}$. Simply replace $\alpha$ by $\beta$ everywhere in the expression for $H_{11}$.

Thus, we have

$$H_{22} = E_0 + J - K$$

The evaluation of $H_{33}$ goes as follows:

$$H_{33} = \frac{1}{2}\iint d\tau_1 d\tau_2 [1s\alpha^*(1)2s\beta^*(2) - 1s\alpha^*(2)2s\beta^*(1)] \times \left[-\frac{1}{2}\nabla_1^2 - \frac{Z}{r_1} - \frac{1}{2}\nabla_2^2 - \frac{Z}{r_2} + \frac{1}{r_{12}}\right] \times [1s\alpha(1)2s\beta(2) - 1s\alpha(2)2s\beta(1)]$$

$$= \frac{1}{2}\iint d\tau_1 d\tau_2 [1s\alpha^*(1)2s\beta^*(2) - 1s\alpha^*(2)2s\beta^*(1)] \left[E_0 + \frac{1}{r_{12}}\right][1s\alpha(1)2s\beta(2) - 1s\alpha(2)2s\beta(1)]$$

$$= \frac{1}{2}[E_0(1 - 0 - 0 + 1) + J - 0 - 0 + J]$$

$$= E_0 + J$$

The evaluation of $H_{44}$ is very similar; simply interchange $\alpha$ and $\beta$ and see that

$$H_{44} = E_0 + J$$

Let's now evaluate $H_{34}$.

$$H_{34} = \frac{1}{2}\iint d\tau_1 d\tau_2 [1s\alpha^*(1)2s\beta^*(2) - 1s\alpha^*(2)2s\beta^*(1)] \times \left[-\frac{1}{2}\nabla_1^2 - \frac{Z}{r_1} - \frac{1}{2}\nabla_2^2 - \frac{Z}{r_2} + \frac{1}{r_{12}}\right] \times [1s\beta(1)2s\alpha(2) - 1s\beta(2)2s\alpha(1)]$$

$$= \frac{1}{2}\iint d\tau_1 d\tau_2 [1s\alpha^*(1)2s\beta^*(2) - 1s\alpha^*(2)2s\beta^*(1)]$$

$$\times \left[E_0 + \frac{1}{r_{12}}\right][1s\beta(1)2s\alpha(2) - 1s\beta(2)2s\alpha(1)]$$

$$= \frac{1}{2}[E_0(0 - 0 - 0 + 0) + 0 - K - K + 0]$$

$$= - K$$

The quantities $H_{12}$, $H_{13}$, $H_{14}$, $H_{23}$ and $H_{24}$ vanish because of the integrations of the spin functions. The evaluation of the $S_{ij}$ involve manipulations that are simiar to those in Problems 8-15 and 8-17. All the off-diagonal $S_{ij}$ ($i \neq j$) vanish because of integrations of the spin functions and the $S_{ii} = 1$ for $i = 1$, 2, 3, and 4.

Putting all these results together, then, gives the secular determinantal equation given in Problem 8-23. The determinant may be evaluated by expanding in minors.

$$D = (E_0 + J - K - E)\begin{vmatrix} E_0+J-K-E & 0 & 0 \\ 0 & E_0+J-E & -K \\ 0 & -K & E_0+J-E \end{vmatrix}$$

$$= (E_0 + J - K - E)^2 \begin{vmatrix} E_0+J-E & -K \\ -K & E_0+J-E \end{vmatrix}$$

$$= (E_0 + J - K - E)^2[(E_0 + J - E)^2 - K^2] = 0$$

The four roots are

$$E = \begin{cases} E_0 + J - K & \text{(twice)} \\ E_0 + J \pm K \end{cases}$$

The $2 \times 2$ block in the secular determinant corresponds to the algebraic equations

$$c_3(E_0 + J - E) - c_4 K = 0$$

$$- c_3 K + c_4(E_0 + J - E) = 0$$

Substitute $E = E_0 + J - K$ into either of these equations and find that

$$c_3 = c_4$$

Thus, the wave function that corresponds to $E = E_0 + J - K$ is

$$\psi_4 = \frac{1}{\sqrt{2}}(\phi_3 + \phi_4)$$

where the factor of $\sqrt{2}$ is a normalization constant.

If we substitute $E = E_0 + J + K$ into either of the above algebraic equations, then we find that

$$c_3 = -\,c_4$$

and that the wave function that corresponds to

$$E = E_0 + J + K$$

is

$$\psi_3 = \frac{1}{\sqrt{2}}(\phi_3 - \phi_4)$$

where, once again, the factor of $\sqrt{2}$ is a normalization constant.

Furthermore,

$$\begin{aligned}\psi_3 &= \frac{1}{2}[1s\alpha(1)2s\beta(2) - 1s\alpha(2)2s\beta(1) \\ &\qquad + 1s\beta(1)2s\alpha(2) - 1s\beta(2)2s\alpha(1)] \\ &= \frac{1}{2}[1s(1)2s(2) - 1s(2)2s(1)] \\ &\qquad \times [\alpha(1)\beta(2) + \beta(1)\alpha(2)]\end{aligned}$$

and

$$\begin{aligned}\psi_4 &= \frac{1}{2}[1s(1)2s(2) + 1s(2)2s(1)] \\ &\qquad \times [\alpha(1)\beta(2) - \beta(1)\alpha(2)]\end{aligned}$$

Following the results of Problem 8-22, we can group the four wave functions $\psi_1$, $\psi_2$, $\psi_3$ and $\psi_4$ into a singlet state ($\psi_4$) and a triplet state ($\psi_1$, $\psi_2$ and $\psi_3$). The energy of the triplet state can be calculated using any of $\psi_1$, $\psi_2$ or $\psi_3$.

$$E(\text{triplet}) = \iint d\tau_1 d\tau_2 \psi_1^* \hat{H} \psi_1$$

$$= \frac{1}{2}\iint d\tau_1 d\tau_2 [1s(1)2s(2) - 1s(2)2s(1)] \times [\alpha^*(1)\alpha^*(2)] \times \left[-\frac{1}{2}\nabla_1^2 - \frac{Z}{r_1} - \frac{1}{2}\nabla_2^2 - \frac{Z}{r_2} + \frac{1}{r_{12}}\right] \times [1s(1)2s(2) - 1s(2)2s(1)][\alpha(1)\alpha(2)]$$

$$= \frac{1}{2}\iint dr_1 dr_2 [1s(1)2s(2) - 1s(2)2s(1)] \left[E_0 + \frac{1}{r_{12}}\right][1s(1)2s(2) - 1s(2)2s(1)]$$

$$= \frac{1}{2}[E_0(1 - 0 - 0 + 1) + J - K - K + J]$$

$$= E_0 + J - K$$

The energy of the singlet state ($\psi_4$) is calculated in a very similar manner and is

$$E(\text{singlet}) = E_0 + J + K$$

(Compare these two results with the four roots of the secular equation).

The integrand in J is everywhere positive, and so $J > 0$. Given that $K > 0$ also, we see that the energy of the triplet state is less than that of the singlet state. Given the values of J and K, we find that

$$E(\text{excited triplet}) = E_0 + J - K = -\frac{5}{2} + \frac{10}{27} - \frac{32}{(27)^2} = -2.174 \text{ au}$$

$$E(\text{excited singlet}) = E_0 + J + K = -\frac{5}{2} + \frac{10}{27} + \frac{32}{(27)^2} = -2.086 \text{ au}$$

The first-order perurbation theory result for the ground-state energy is given in Section 8-2 and Problem 8-6

$$E(\text{ground state}) = -\frac{11}{4} = -2.750 \text{ au}$$

Therefore, we have

$$\Delta E(\text{triplet to ground state}) = -2.174 + 2.750 = 0.576 \text{ au}$$

$$= 15.7 \text{ eV}$$

and

$$\Delta E(\text{singlet to ground state}) = -2.086 + 2.750$$
$$= 0.664 \text{ au}$$
$$= 18.1 \text{ eV}$$

**8-24.** The Slater determinantal wave function for the ground state of a lithium atom is given by Eq. 8-70

$$\Psi = \frac{1}{\sqrt{3!}}\begin{vmatrix} 1s\alpha(1) & 1s\beta(1) & 2s\alpha(1) \\ 1s\alpha(2) & 1s\beta(2) & 2s\alpha(2) \\ 1s\alpha(3) & 1s\beta(3) & 2s\alpha(3) \end{vmatrix}$$

$$= \frac{1}{\sqrt{3!}}\Big[1s(1)1s(2)2s(3) - 2s(1)1s(2)1s(3)\Big]\alpha(1)\beta(2)\alpha(3)$$
$$+ \frac{1}{\sqrt{3!}}\Big[2s(1)1s(2)1s(3) - 1s(1)2s(2)1s(3)\Big]\alpha(1)\alpha(2)\beta(3)$$
$$+ \frac{1}{\sqrt{3!}}\Big[1s(1)2s(2)1s(3) - 1s(1)1s(2)2s(3)\Big]\beta(1)\alpha(2)\alpha(3)$$

Each of the products of spin functions is an eigenfunction of $\hat{S}_{z,\text{total}} = \hat{S}_{z1} + \hat{S}_{z2} + \hat{S}_{z3}$ with eigenvalue $+\hbar/2$ (see, for example, Problems 8-20 and 8-21), and so $\Psi$ is an eigenfunction of $\hat{S}_{z,\text{total}}$ with eigenvalue $+\hbar/2$, or 1/2 in atomic units.

**8-25.** This problem is more difficult than most of the others. To do this problem, you must first realize that a Slater determinant can be written as

$$\Psi(1,2,\ldots N) = (N!)^{-1/2}\sum_P \varepsilon_P u_1(P1)u_2(P2)\ldots u_N(PN)$$

where P represents a permutation of the numbers (1,2,3,...,N) and $\varepsilon_P = +$ if the permutation (P1, P2, ...,PN) differs from (1, 2, ...,N) by an even number of interchanges of pairs of numbers or

$\varepsilon_P = -1$ if the permutation (P1, P2, ...,PN) differs from (1,2,...N) by an odd number of interchanges. An equivalent way of writing the above formula is

$$\Psi(1,2,\ldots N) = (N!)^{-1/2}\sum_P \varepsilon_P u_{P1}(1)u_{P2}(2)\ldots u_{PN}(N)$$

In other words, we interchange the labels of the spin-orbitals rather than the electrons. We shall use both expressions for $\Psi$ in this problem. First, let's use this notation to show that $\Psi$ is normalized. We have

$$\int\cdots\int\Psi^*\Psi d\tau_1 d\tau_2\cdots d\tau_N$$

$$= (N!)^{-1}\sum_P \varepsilon_P \sum_Q \varepsilon_Q \int\cdots\int u_1^*(Q1)u_2^*(Q2)\cdots$$
$$\times\, u_N(QN)u_1(P1)u_2(P2)\cdots u_N(PN)d\tau_1\cdots d\tau_N$$

These integrals will vanish unless Q1 = P1, Q2 = P2, and so on. This means that only permutation Q that will lead to a non-zero result is when Q = P. This leaves

$$\int\cdots\int\Psi^*\Psi d\tau_1\cdots d\tau_N$$

$$= (N!)^{-1}\sum_P \varepsilon_P^2\int\cdots\int u_1^*(P1)u_1(P1)u_2^*(P2)u_2(P2)$$
$$\cdots u_N^*(PN)u_N(PN)d\tau_1\cdots d\tau_N$$

where $\varepsilon_P^2 = 1$.

Because the integration variables are dummy variables, all these integrals are equal, and equal to unity because the individual spin-orbitals are normalized. There are N! permutations, and so

$$\int\cdots\int\Psi^*\Psi d\tau_1\cdots d\tau_N = 1$$

Now let's consider integrals of the type

$$A_1 = \int\cdots\int\Psi^*\left[\sum_j\hat{H}(j)\right]\Psi d\tau_1\cdots d\tau_N$$

where $\hat{H}(j)$ is a so-called one-electron operator,

meaning that it operates on the coordinates of electron j only.

For example,

$$\sum_j \hat{H}(j) = \sum_j \left[-\frac{1}{2}\nabla_j^2 - \frac{Z}{r_j}\right]$$

in Eq. 8-71.

Using the above expansion for $\Psi$, we write

$$A_1 = (N!)^{-1}\sum_P \varepsilon_P \sum_Q \varepsilon_Q \int\cdots\int u_1^*(Q1)u_2^*(Q2)\cdots$$
$$\times\, u_N^*(QN)\left[\sum_j \hat{H}(j)\right]u_1(P1)u_2(P2)\cdots$$
$$\times\, u_N(PN)d\tau_1\cdots d\tau_N$$

This integral will vanish unless Q1 = P1, Q2 = P2, and so on. Thus, the only permutation Q that gives a non-zero result is Q = P, and so

$$A_1 = (N!)^{-1}\sum_P \varepsilon_P^2 \int\cdots\int u_1^*(P1)\cdots$$
$$u_N^*(PN)\left[\sum_j \hat{H}(j)\right]u_1(P1)\cdots u_N(PN)d\tau_1\cdots d\tau_N$$

As above, all N! permutations lead to the same integral, and so

$$A_1 = \sum_j \int u_j^*(j)\hat{H}(j)u_j(j)d\tau_j$$

Finally, let's consider integrals of the type

$$A_2 = \int\cdots\int \Psi^*\left[\sum_{i<j}\frac{1}{r_{ij}}\right]\Psi d\tau_1\cdots d\tau_N$$
$$= (N!)^{-1}\sum_P \varepsilon_P \sum_Q \varepsilon_Q \int\cdots\int u_{Q1}^*(1)u_{Q2}^*(2)\cdots$$
$$\times\, u_{QN}^*(QN)\left[\sum_{i<j}\frac{1}{r_{ij}}\right]u_{P1}(1)u_{P2}(2)\cdots$$
$$\times\, u_{PN}(N)d\tau_1\cdots d\tau_N$$

These integrals will vanish unless the subscripts of all the spin-orbitals _except_ those for electrons i and j match. There will be two non-vanishing terms in the summation of all Q permutations

$$A_2 = (N!)^{-1}\sum_P \left[\left[\varepsilon_P \sum_{i<j} \iint u_{P1}^*(i)u_{P2}(j) \times \frac{1}{r_{ij}} u_{P1}(i)u_{P2}(j)d\tau_i d\tau_j\right] - \varepsilon_P \sum_{i<j} \iint u_{P2}^*(i)u_{P1}(j)\frac{1}{r_{ij}} u_{P1}(i)u_{P2}(j)d\tau_i d\tau_j\right]$$

Note that Q = P for the term with the positive sign and that Q and P differ by the interchange of one pair of subscripts for the term with the negative sign. As before, all N! permutations lead to the same quantities, and so,

$$A_2 = \sum_{i<j} \iint u_i^*(1)u_j^*(2)\frac{1}{r_{12}} u_i(1)u_j(2)d\tau_1 d\tau_2\Big] - \sum_{i<j} \iint u_i^*(1)u_j^*(2)\frac{1}{r_{12}} u_i(2)u_j(1)d\tau_1 d\tau_2\Big]$$

$$= \sum_{i<j} J'_{ij} - \sum_{i<j} K'_{ij}$$

Note that the integrals here involve spin-orbitals, whereas the integrals in Eqs. 8-76 and 8-77 involve only spatial orbitals.

Putting everything together, then, we have

$$E = \int\cdots\int \Psi^*\hat{H}\Psi d\tau_1\cdots d\tau_N$$

$$= \sum_j I_j + \sum_{i<j} \iint u_i^*(1)u_j^*(2)\frac{1}{r_{12}} u_i(1)u_j(2)d\tau_1 d\tau_2 - \sum_{i<j} \iint u_i^*(1)u_j^*(2)\frac{1}{r_{12}} u_i(2)u_j(1)d\tau_1 d\tau_2$$

$$= \sum_j I_j + \sum_{i<j} J'_{ij} - \sum_{i<j} K'_{ij}$$

where $I_j$ is defined by Eq. 8-75.

If the Slater orbital is of the form of Eq. 8-72, namely N doubly-occupied orbitals, then $I_1 = I_2$, $I_3 = I_4$, and so forth.

Furthermore,

$$J'_{12} = J_{11},\ J'_{13} = J_{12},\ J'_{14} = J_{12}$$

$$J'_{23} = J_{12},\ J'_{34} = J_{22}$$

and so forth. Finally,

$$K'_{12} = 0,\ K'_{13} = K_{12},\ K'_{14} = 0$$

and so forth, giving

$$E = 2\sum_{j=1}^{N} I_j + \sum_{i=1}^{N}\sum_{j=1}^{N}(2J_{ij} - K_{ij})$$

where the summations now are over spatial orbitals. If you do not see the relations between the $J'_{ij}$ and the $J_{ij}$ and the $K'_{ij}$ and the $K_{ij}$, take a beryllium atom as a concrete example.

**8-26.** The integrands of the $J_{ij}$ (Eq. 8-76) are of the form

$$\phi_i^*(\mathbf{r}_1)\phi_i(\mathbf{r}_1)\frac{1}{r_{12}}\phi_j^*(\mathbf{r}_2)\phi_j(\mathbf{r}_2)d\mathbf{r}_1 d\mathbf{r}_2$$

which can be interpreted as the electrostatic, coulombic interaction betwen two charges of magnitudes $\phi_i^*(\mathbf{r}_1)\phi_i(\mathbf{r}_1)d\mathbf{r}_1$ and $\phi_j^*(\mathbf{r}_2)\phi_j(\mathbf{r}_2)d\mathbf{r}_2$ separated by a distance $r_{12}$. The integrands of the $K_{ij}$ ($i \neq j$) do not lend themselves to such a classical interpretation. The $K_{ij}$ may be obtained from the $J_{ij}$ by exchanging $\mathbf{r}_1$ and $\mathbf{r}_2$ in either $\phi_i^*(\mathbf{r}_1)\phi_j^*(\mathbf{r}_2)$ or $\phi_i(\mathbf{r}_1)\phi_j(\mathbf{r}_2)$.

**8-27.** Start with Eq. 8-74

$$E = 2\sum_{j=1}^{N} I_j + \sum_{i=1}^{N}\sum_{j=1}^{N}(2J_{ij} - K_{ij})$$

which is a summation over doubly-occupied orbitals. Let's remove an electron from the Nth orbital. Then the energy is

$$E^+ = 2\sum_{j=1}^{N-1} I_j + I_N + \sum_{i=1}^{N-1}\sum_{j=1}^{N-1}(2J_{ij} - K_{ij})$$

The difference between E and $E^+$ is

$$E - E^+ = I_N + \sum_{i=1}^{N}(2J_{iN} - K_{iN})$$

Now multiply Eq. 8-78 by $\phi_i^*$ and integrate to obtain

$$\varepsilon_i = I_i + 2\sum_j J_{ij} - \sum_j K_{ij}$$

Note that $\varepsilon_N$ is equal to $E - E^+$. If we had removed an electron from the ith orbital instead of the Nth orbital, we would have obtained

$$\varepsilon_i = E - E^+$$

To make these equations more concrete, let's consider Be and $Be^+$. Beryllium has four electrons, and its ground-state electron configuration is $1s^2 2s^2$. Thus

$$E = 2I_1 + 2I_2 + (2J_{11} - K_{11})$$
$$+ (2J_{22} - K_{22}) + 2(2J_{12} - K_{12})$$
$$= 2I_1 + 2I_2 + J_{11} + J_{22} + 4J_{12} - 2K_{12}$$

where we have used the fact that $J_{ii} = K_{ii}$. The ground-state electron configuration of $Be^+$ is $1s^2 2s^1$, and

$$E^+ = 2I_1 + I_2 + J_{11} + 2J_{12} - K_{12}$$

because one of the 2s electrons has been removed from Be. Thus, we have

$$E - E^+ = I_2 + J_{22} + 2J_{12} - K_{12}$$
$$= I_2 + (2J_{12} - K_{12}) + (2J_{22} - K_{22})$$

But this result is the same as

$$\varepsilon_2 = I_2 + 2\sum_j J_{2j} - \sum_j K_{2j}$$
$$= I_2 + (2J_{12} - K_{12}) + (2J_{22} - K_{22})$$

**8-28.** If we multiply Eq. 8-78 by $\phi_i^*(\mathbf{r}_1)$ and integrate over $\mathbf{r}_1$, then we obtain

$$\varepsilon_i = \int d\mathbf{r}_1 \phi_i^*(\mathbf{r}_1) \hat{F}_i \phi_i(\mathbf{r}_1)$$

$$= \int d\mathbf{r}_1 \phi_i^*(\mathbf{r}_1) \left[\hat{f}_i + \sum_j (2\hat{J}_j - \hat{K}_j)\right] \phi_i(\mathbf{r}_1)$$

$$= I_i + \sum_j (2J_{ij} - K_{ij})$$

Now

$$\sum_{i=1}^{N} (\varepsilon_i + I_i) = 2\sum_{i=1}^{N} I_i + \sum_{i=1}^{N} \sum_{j=1}^{N} (2J_{ij} - K_{ij})$$
$$= E$$

**8-29.** For an $np^1$ electron configuration, there are six entries in a table of possible sets of m and $m_s$. We have

| m | $m_s$ | M | $M_S$ | $M_J$ |
|---|---|---|---|---|
| 1 | +1/2 | 1 | +1/2 | 3/2 |
| 1 | −1/2 | 1 | −1/2 | 1/2 |
| 0 | +1/2 | 0 | +1/2 | 1/2 |
| 0 | −1/2 | 0 | −1/2 | −1/2 |
| −1 | +1/2 | −1 | +1/2 | −1/2 |
| −1 | −1/2 | −1 | −1/2 | −3/2 |

The M and $M_S$ values in this table correspond to a $^2P$ state. The $M_J$ values consist of two sets:
3/2, 1/2, −1/2, −3/2 and 1/2, −1/2, corresponding to J = 1/2 and 3/2. Thus, the term symbols associated with an $np^1$ electron configuration are $^2P_{3/2}$ and $^2P_{1/2}$. The ground state is determined by applying the third of Hund's rules (p. 319). The ground state is $^2P_{1/2}$.
The term symbols asociated with an $np^5$ electron configuration are the same as for an $np^1$ electron configuration. There are six entries in the table (see Eq. 8-89), and it is like having one electron "hole" to be filled.

**8-30.** There are 15 entries in a table of possible m and $m_s$ values for a $np^4$ electron configuration, just as for an $np^2$ electron configuration. In deducing the term symbols for an $np^4$ electron configuration, we have two electron "holes" to fill, giving the same table of entries in having two spin orbitals to fill as in an $np^2$ electron configuration.

**8-31.** Each value of M associated with a value of L has 2S + 1 entries, and since there are 2L + 1 values of M for each value of L, the total number of entries for each term symbol (excluding the J subscript) is (2L + 1)(2S + 1). Applying this result to the $np^2$ case gives

$$\underset{^1S}{1 \times 1} + \underset{^3P}{(3 \times 3)} + \underset{^1D}{(1 \times 5)} = 15$$

**8-32.** There are (10!/8!2!) = 45 entries in a table of possible values of m and $m_s$. The number of entries accounted for by the term symbols given in the problem are

$$\underset{^1S}{(1 \times 1)} + \underset{^1D}{(1 \times 5)} + \underset{^3P}{(3 \times 3)} + \underset{^3F}{(3 \times 7)} + \underset{^1G}{(1 \times 9)}$$
$$= 45$$

**8-33.** There are 2 × 6 = 12 possible sets of m and $m_s$.

| | $m_1$ | $m_{1s}$ | $m_2$ | $m_{2s}$ | M | $M_S$ | $M_J$ |
|---|---|---|---|---|---|---|---|
| 1. | 0 | +1/2 | 1 | +1/2 | 1 | 1 | 2 |
| 2. | 0 | −1/2 | 1 | +1/2 | 1 | 0 | 1 |
| 3. | 0 | +1/2 | 1 | −1/2 | 1 | 0 | 1 |
| 4. | 0 | −1/2 | 1 | −1/2 | 1 | −1 | 0 |
| 5. | 0 | +1/2 | 0 | +1/2 | 0 | 1 | 1 |
| 6. | 0 | −1/2 | 0 | +1/2 | 0 | 0 | 0 |
| 7. | 0 | +1/2 | 0 | −1/2 | 0 | 0 | 0 |
| 8. | 0 | −1/2 | 0 | −1/2 | 0 | −1 | −1 |
| 9. | 0 | +1/2 | −1 | +1/2 | −1 | 1 | 0 |
| 10. | 0 | −1/2 | −1 | +1/2 | −1 | 0 | −1 |
| 11. | 0 | +1/2 | −1 | −1/2 | −1 | 0 | −1 |
| 12. | 0 | −1/2 | −1 | −1/2 | −1 | −1 | −2 |

Entries 1, 2, 4, 5, 6, 8, 9, 10 and 12 correspond to L = 1 and S = 1, or a $^3P$ term symbol. Entries 3, 7 and 11 correspond to L = 1 and S = 0, yielding $^1P$. The values of J can be deduced from the table, or from Eq. 8-90. The final result is

| $^3P_2$ | $^3P_1$ | $^3P_0$ | $^1P_1$ |
|---|---|---|---|
| (L + S) | (L + S − 1) | (\|L − S\|) | (L + S) |

According to Hund's rules, the ground state is $^3P_0$.

**8-34.** There are 2 × 10 = 20 sets of m and $m_s$ for an $ns^1nd^1$ electron configuration. The determination of the term symbols is a direct extension of Problem 8-33, and the term symbols are $^3D$ and $^1D$. The values of J can be determined from Eq. 8-90, and the final result is

| $^3D_3$ | $^3D_2$ | $^3D_1$ | $^1D_2$ |
|---|---|---|---|
| (L + S) | (L + S − 1) | (\|L − S\|) | L + S |

According to Hund's rules, the ground state is $^3D_1$.

**8-35.** The values of J can be determined from Eq. 8-90. Let's consider each term symbol in turn.

$^1S$: L = 0, S = 0, J = 0 $\qquad ^1S_0$

$^1D$: L = 2, S = 0, J = 2 $\qquad ^1D_2$

$^3P$: $L = 1, S = 1, J = L + S = 2$ $\quad {}^3P_2$

$= L + S - 1 = 1$ $\quad {}^3P_1$

$= |L - S| = 0$ $\quad {}^3P_0$

$^3F$: $L = 3, S = 1, J = L + S = 4$ $\quad {}^3F_4$

$= L + S - 1 = 3$ $\quad {}^3F_3$

$= |L - S| = 2$ $\quad {}^3F_2$

$^1G$: $L = 4, S = 0, J = 4$ $\quad {}^1G_4$

**8-36.** The values of J can be determined from Eq. 8-90. Let's consider each term symbol in turn.

$^2P$: $L = 1,\ S = \frac{1}{2},\ J = L + S = \frac{3}{2}$ $\quad {}^2P_{3/2}$

$= |L - S| = \frac{1}{2}$ $\quad {}^2P_{1/2}$

$^2D$: $L = 2,\ S = \frac{1}{2},\ J = L + S = \frac{5}{2}$ $\quad {}^2D_{5/2}$

$= |L - S| = \frac{3}{2}$ $\quad {}^2D_{3/2}$

$^4S$: $L = 0,\ S = \frac{3}{2},\ J = \frac{3}{2}$ $\quad {}^4S_{3/2}$

According to Hund's rules, the ground state is $^4S_{3/2}$.

**8-37.** The ground-state electron configuration of a magnesium atom is $1s^22s^22p^63s^2$, or $[Ne]3s^2$. From pp. 314-315, the term symbol for a $3s^2$ electron configuration is $^1S_0$.

**8-38.** The term symbols for an $nd^2$ electron configuration are given in Problem 8-35. They are

$^1S_0 \quad {}^1D_2 \quad {}^1G_4 \quad {}^3P_2 \quad {}^3P_1$

$^3P_0 \quad {}^3F_4 \quad {}^3F_3 \quad {}^3F_2$

According to Hund's rules, the ground state is $^3F_2$.

**8-39.** The electron configuration is $[Kr]4d^{10}$. Because all the subshells are filled, the term symbol is $^1S_0$.

**8-40.** The Lyman series of atomic hydrogen consists of $n \rightarrow 1$ transitions. The selection rule $\Delta L = \pm 1$ (Eq. 8-93) means that we consider only P to S transitions. The separation of the doublets can be obtained by

$$\begin{aligned} E(2p\ ^2P_{3/2}) - E(2p\ ^2P_{1/2}) &= 82{,}259.272\ \text{cm}^{-1} - 82{,}258.917\ \text{cm}^{-1} \\ &= 0.355\ \text{cm}^{-1} \end{aligned}$$

$$\begin{aligned} E(3p\ ^2P_{3/2}) - E(3p\ ^2P_{1/2}) &= 97{,}492.306\ \text{cm}^{-1} - 97{,}492.198\ \text{cm}^{-1} \\ &= 0.108\ \text{cm}^{-1} \end{aligned}$$

$$\begin{aligned} E(4p\ ^2P_{3/2}) - E(4p\ ^2P_{1/2}) &= 102{,}823.881\ \text{cm}^{-1} - 102{,}823.835\ \text{cm}^{-1} \\ &= 0.046\ \text{cm}^{-1} \end{aligned}$$

**8-41.** From Table 8-7,

$$\begin{aligned} \Delta E = E(4f\ ^2F) - E(3d\ ^2D) &= 34{,}588.6\ \text{cm}^{-1} - 29{,}172.9\ \text{cm}^{-1} \\ &= 5415.7\ \text{cm}^{-1} \end{aligned}$$

The wavelength (in vacuum) is

$$\lambda_{vac} = \frac{1}{\Delta E} = 1.8465 \times 10^{-4}\ \text{cm} = 18465\ \text{Å}$$

The data in Figure 8-5 give $\lambda_{air}$, and so

$$\lambda_{air} = \frac{\lambda_{vac}}{1.00029} = 18459\ \text{Å}$$

in excellent agreement with Figure 8-5.

**8-42.**

| sharp series | principal series |
|---|---|
| 11 404 Å | 5 895.9 Å |
| 11 382 Å | 5 889.9 Å |
| 6 160.7 Å | 3 302.9 Å |
| 6 154.2 Å | 3 302.3 Å |
| 5 153.6 Å | 2 853.0 Å |
| 5 149.1 Å | 2 852.8 Å |

| diffuse series | fundamental series |
|---|---|
| 8 194.8 Å | 18 459 Å |
| 8 183.3 Å | |
| 5 688.2 Å | 12 678 Å |
| 5 682.7 Å | |
| 4 982.9 Å | |
| 4 978.6 Å | |

**8-43.** Values of $\bar{\nu}_{air}$ from Table 8-7. (For the diffuse and fundamental series, we have ignored the splitting in the nd and nf levels.)

| sharp series | principal series |
|---|---|
| 8 766.48 $cm^{-1}$ | 16 956.183 $cm^{-1}$ |
| 8 783.68 $cm^{-1}$ | 16 973.379 $cm^{-1}$ |
| 16 227.317 $cm^{-1}$ | 30 266.88 $cm^{-1}$ |
| 16 244.513 $cm^{-1}$ | 30 272.51 $cm^{-1}$ |
| 19 399.268 $cm^{-1}$ | 35 040.27 $cm^{-1}$ |
| 19 416.464 $cm^{-1}$ | 35 042.79 $cm^{-1}$ |

| diffuse series | fundamental series |
|---|---|
| 12 199.5 $cm^{-1}$ | 5 415.7 $cm^{-1}$ |
| 12 216.7 $cm^{-1}$ | |
| 17 575.4 $cm^{-1}$ | 7 884.7 $cm^{-1}$ |
| 17 592.6 $cm^{-1}$ | |
| 20 063.4 $cm^{-1}$ | |
| 20 080.6 $cm^{-1}$ | |

We now convert these results to $\lambda_{air}$ by using the formula

$$\lambda_{air} = \frac{1}{1.00029\bar{\nu}_{vac}}$$

| sharp series | principal series |
|---|---|
| 11 404 Å | 5 895.9 Å |
| 11 381 Å | 5 889.9 Å |
| 6 160.7 Å | 3 303.0 Å |
| 6 154.1 Å | 3 302.4 Å |
| 5 153.4 Å | 2 853.0 Å |
| 5 148.8 Å | 2 852.8 Å |

| diffuse series | fundamental series |
|---|---|
| 8 194.7 Å | 18 459 Å |
| 8 183.2 Å | |
| 5 688.1 Å | 12 679 Å |
| 5 682.6 Å | |
| 4 982.8 Å | |
| 4 978.5 Å | |

**8-44.**

$$E^{(1)} = \iint d\mathbf{r}_1 d\mathbf{r}_2 \psi_{1s}(\mathbf{r}_1)\psi_{1s}(\mathbf{r}_2)\frac{1}{r_{12}}\psi_{1s}(\mathbf{r}_1)\psi_{1s}(\mathbf{r}_2)$$

$$= \frac{Z^6}{\pi^2}\int_0^\infty dr_1 r_1^2 \exp(-2Zr_1)\int_0^\infty dr_2 r_2^2 \exp(-2Zr_2)\int_0^{2\pi} d\phi_1 \int_0^{2\pi} d\phi_2$$

$$\times \int_0^\pi d\theta_1 \sin\theta_1 \int_0^\pi d\theta_2 \sin\theta_2$$

$$\times \left[\sum_{\ell=0}^{\infty}\sum_{m=-\ell}^{+\ell}\frac{4\pi}{2\ell+1}\frac{r_<^\ell}{r_>^{\ell+1}}Y_\ell^m(\theta_1,\phi_1)Y_\ell^m(\theta_2,\phi_2)^*\right]$$

The integrals over the angles can be written as

$$\left[\int_0^{2\pi} d\phi_1 \int_0^\pi d\theta_1 \sin\theta_1 Y_\ell^m(\theta,\phi)\right]\left[\int_0^{2\pi} d\phi_2 \int_0^\pi d\theta_1 \sin\theta_1 Y_\ell^m(\theta,\phi)^*\right]$$

Both of these integrals vanish unless $\ell = m = 0$ because of the orthogonality relation (Eq. 6-77)

$$\int_0^{2\pi} d\phi \int_0^\pi d\theta \sin\theta Y_\ell^m(\theta,\phi)Y_0^0(\theta,\phi)^* = \delta_{\ell 0}\delta_{m0}$$

where $Y_0^0(\theta,\phi)$ is $(4\pi)^{-1/2}$.

When $\ell = m = 0$, each of the above integrals equals $4\pi$ and $E^{(1)}$ becomes

$$E^{(1)} = 16Z^6\int_0^\infty dr_1 r_1^2 \exp(-2Zr_1) \int_0^\infty dr_2 r_2^2 \frac{\exp(-2Zr_2)}{r_>}$$

We now break up the integral over $r_2$ into two pieces, $r_2 < r_1$ and $r_2 > r_1$. When $r_2 < r_1$, $r_> = r_1$ and when $r_2 > r_1$, $r_> = r_2$. Thus, $E^{(1)}$ becomes

$$E^{(1)} = 16Z^6\int_0^\infty dr_1 r_1 \exp(-2Zr_1) \int_0^{r_1} dr_2 r_2^2 \exp(-2Zr_2)$$

$$+ 16Z^6\int_0^\infty dr_1 r_1^2 \exp(-2Zr_1) \int_{r_1}^\infty dr_2 r_2 \exp(-2Zr_2)$$

$$= 4Z^3\int_0^\infty dr_1 r_1 \exp(-2Zr_1) \times \left[1 - [\exp(-2Zr_1) \times (2Z^2r_1^2 + 2Zr_1 + 1)]\right]$$

$$+ 4Z^4\int_0^\infty dr_1 r_1^2 \exp(-2Zr_1) \times [\exp(-2Zr_1)(2Zr_1 + 1)]$$

In evaluating these integrals over $r_1$, it is convenient to let $x = 2Zr_1$

$$E^{(1)} = Z\int_0^\infty dx\ xe^{-x}\left[1 - e^{-x}\left[\frac{1}{2}x^2 + x + 1\right]\right]$$

$$+ \frac{Z}{2}\int_0^\infty dx\ x^2 e^{-x}[e^{-x}(x + 1)]$$

$$= Z\int_0^\infty dx\ xe^{-x} - \frac{Z}{2}\int_0^\infty dx e^{-2x}(x^2 + 2x)$$

$$= Z - \frac{Z}{2}\left[\frac{2!}{8} + \frac{2}{4}\right] = \frac{5}{8}Z$$

# CHAPTER NINE

# MOLECULES

**9-1.** Start with Eq. 9-2 and let $\hbar = 1$, $m = 1$, $e = 1$, $4\pi\varepsilon_0 = 1$ and $Z = 1$ and obtain Eq. 9-3.

**9-2.** This problem is done on pp. 267 and 268 (see Eqs. 7-54 to 7-59).

**9-3.** From Figure 9-36, the sum $r_A + r_B$ can never be less than R, and so $\lambda \geq 1$. The difference $r_A - r_B$ can never exceed $\pm R$ (colinear case), and so $-1 \leq \mu \leq +1$. To evaluate the overlap integral S

$$S = \frac{R^3Z^3}{8\pi}\int_0^{2\pi} d\phi \int_1^{\infty} d\lambda \int_{-1}^{1} d\mu(\lambda^2 - \mu^2)e^{-Zr_A}\, e^{-Zr_B}$$

$$= \frac{R^3Z^3}{4}\int_1^{\infty} d\lambda \int_{-1}^{1} d\mu(\lambda^2 - \mu^2)e^{-ZR\lambda}$$

$$= \frac{R^3Z^3}{4}\int_1^{\infty} d\lambda\; e^{-ZR\lambda} \int_{-1}^{1} d\mu(\lambda^2 - \mu^2)$$

$$= \frac{R^3Z^3}{4}\int_1^{\infty} d\lambda\; e^{-ZR\lambda}\left[2\lambda^2 - \frac{2}{3}\right]$$

$$= \frac{R^3Z^3}{2}\left[\left[\frac{1}{ZR} + \frac{2}{Z^2R^2} + \frac{2}{Z^3R^3}\right]e^{-ZR} - \frac{1}{3ZR}e^{-ZR}\right]$$

$$= e^{-ZR}\left[1 + ZR + \frac{Z^2R^2}{3}\right]$$

**9-4.**

$$J_2 = \frac{R^3}{8\pi}\int_0^{2\pi} d\phi \int_1^{\infty} d\lambda \int_{-1}^{1} d\mu(\lambda^2 - \mu^2)\frac{e^{-R(\lambda+\mu)}}{R(\lambda - \mu)/2}$$

where we have used the relations $2r_{1A} = R(\lambda + \mu)$ and $2r_{1B} = R(\lambda - \mu)/2$. Thus

$$J_2 = \frac{R^2}{2}\int_1^{\infty} d\lambda e^{-R\lambda} \int_{-1}^{1} d\mu\frac{(\lambda^2 - \mu^2)}{(\lambda - \mu)}e^{-R\mu}$$

$$= \frac{R^2}{2}\int_1^\infty d\lambda e^{-R\lambda} \int_{-1}^{1} d\mu(\lambda + \mu) e^{-R\mu}$$

$$= \frac{R^2}{2}\int_1^\infty d\lambda e^{-R\lambda} \left[\lambda\left[\frac{e^R - e^{-R}}{R}\right] + \left[\frac{e^R(1 - R)}{R^2} - \frac{e^{-R}(1 + R)}{R^2}\right]\right.$$

$$= \frac{R}{2}(e^R - e^{-R})\int_1^\infty d\lambda\ \lambda e^{-R\lambda} + \left[\frac{(e^R - e^{-R})}{2} - \frac{R(e^R + e^{-R})}{2}\right]\int_1^\infty d\lambda e^{-R\lambda}$$

$$= \frac{R(e^R - e^{-R})}{2}\left[\frac{e^{-R}(1 + R)}{R^2}\right] + \left[\frac{(e^R - e^{-R})}{2} - \frac{R(e^R + e^{-R})}{2}\right]\left[\frac{e^{-R}}{R}\right]$$

$$= \frac{(1 - e^{-2R})}{2}\left[1 + \frac{1}{R}\right] + \frac{(1 - e^{-2R})}{2R} - \frac{(1 + e^{-2R})}{2}$$

$$= \frac{1}{R} - e^{-2R}\left[1 + \frac{1}{R}\right]$$

To evaluate $J_1$, we write

$$J_1 = \frac{R^3}{8\pi}\int_0^{2\pi} d\phi \int_1^\infty d\lambda \int_{-1}^{1} d\mu(\lambda^2 - \mu^2)\exp(-2R_{2B}) \times \left[\frac{1}{r_{2A}} - \exp(-2R_{2A})\left[1 + \frac{1}{r_{2A}}\right]\right]$$

$$= \frac{R^3}{4}\int_1^\infty d\lambda \int_{-1}^{1} d\mu(\lambda^2 - \mu^2)\left[\frac{e^{-R(\lambda-\mu)}}{R(\lambda + \mu)/2} - e^{-2R\lambda}\left[1 + \frac{1}{R(\lambda + \mu)/2}\right]\right]$$

$$= \frac{R^3}{4}\int_1^\infty d\lambda \int_{-1}^{1} d\mu\left[\frac{2}{R}(\lambda - \mu)\ e^{-R\lambda}\ e^{R\mu} - e^{-2R\lambda}\left[\lambda^2 - \mu^2 + \frac{2}{R}(\lambda - \mu)\right]\right]$$

$$= \frac{R^2}{2}\int_1^\infty d\lambda e^{-R\lambda} \int_{-1}^{1} d\mu(\lambda - \mu)e^{R\mu}$$

$$- \frac{R^3}{4}\int_1^\infty d\lambda e^{-2R\lambda} \int_{-1}^{1} d\mu\left[\lambda^2 - \mu^2 + \frac{2(\lambda - \mu)}{R}\right]$$

$$= \frac{R^2}{2}\int_1^\infty d\lambda e^{-R\lambda}\left[\lambda\frac{(e^R - e^{-R})}{R} - \frac{e^R(R - 1) + e^{-R}(R + 1)}{R^2}\right]$$

$$- \frac{R^3}{4}\int_1^\infty d\lambda e^{-2R\lambda}\left[2\lambda^2 - \frac{2}{3} + \frac{4\lambda}{R}\right]$$

$$= \frac{R(e^R - e^{-R})}{2}\left[\frac{e^{-R}(R + 1)}{R^2}\right]$$

$$- \left[\frac{e^R(R - 1) + e^{-R}(R + 1)}{2}\right]\left[\frac{e^{-R}}{R}\right]$$

$$- \frac{R^3}{4}\left[2e^{-2R}\left[\frac{1}{2R} + \frac{1}{2R^2} + \frac{1}{4R^3}\right]\right.$$

$$\left. - \frac{2}{3}\left[\frac{e^{-2R}}{2R}\right] + \frac{4}{R}\left[\frac{e^{-2R}(1 + 2R)}{4R^2}\right]\right]$$

$$= \frac{R + 1}{2R} - \frac{(R - 1)}{2R} - e^{-2R}\left[\frac{R + 1}{2R} + \frac{R + 1}{2R}\right.$$

$$\left. + \frac{R^2}{4} + \frac{R}{4} + \frac{1}{8} - \frac{R^2}{12} + \frac{(1 + 2R)}{4}\right]$$

$$= \frac{1}{R} - e^{-2R}\left[\frac{1}{R} + \frac{11}{8} + \frac{3R}{4} + \frac{R^2}{6}\right]$$

**9-5.**

$$H_{11} = \iint d\mathbf{r}_1 d\mathbf{r}_2 1s_A(1)1s_B(2)\ \hat{H}1s_A(1)1s_B(2)$$

$$H_{22} = \iint d\mathbf{r}_1 d\mathbf{r}_2 1s_A(2)1s_B(1)\ \hat{H}1s_A(2)1s_B(1)$$

The Hamiltonian is symmetric in the indices 1 and 2, and so $H_{22}$ becomes equal to $H_{11}$ if we simply interchange the integration variables 1 and 2 in $H_{22}$.

**9-6.** Either term separately in a Heitler-London wave function for $H_2$ will lead to $H_{11} = H_{22} = E$.

**9-7.** Start with Eq. 9-33

$$c_1(-1 + J - E_+) + c_2(-S^2 + K - S^2E_+) = 0$$

Substitute

$$E_+ = -1 + \frac{J + K}{1 + S^2}$$

into this to obtain

$$c_1\left[\frac{JS^2 - K}{1 + S^2}\right] + c_2\left[\frac{K - S^2J}{1 + S^2}\right] = 0$$

or

$$c_1 = c_2$$

**9-8.** From Eq. 9-32

$$\Delta E_+ = \frac{J + K}{1 + S^2}$$

where

$$K = \iint d\mathbf{r}_1 d\mathbf{r}_2 [1s_A(1)1s_B(1)]$$
$$\times \left[-\frac{1}{r_{1A}} - \frac{1}{r_{2B}} + \frac{1}{r_{12}} + \frac{1}{R}\right][1s_A(2)1s_B(2)]$$
$$= -\int d\mathbf{r}_2[1s_A(2)1s_B(2)] \int d\mathbf{r}_1 \frac{1s_A(1)1s_B(1)}{r_{1A}}$$
$$- \int d\mathbf{r}_1[1s_A(1)1s_B(1)] \int d\mathbf{r}_2 \frac{1s_A(2)1s_B(2)}{r_{1B}}$$
$$+ (ab|ab) + \frac{S^2}{R}$$
$$= -2S(a|ab) + (ab|ab) + S^2/R$$

**9-9.** The observed dissociation energy is the minimum eergy required to dissociate a molecule from its ground state. Because of the zero-point vibrational energy ($h\nu/2$) in the ground state, we have

$$D_o = D_e - \frac{1}{2}h\nu$$

We now expand the exponential function in $u(r)$

$$\exp[-a(r - r_e)]$$
$$= 1 - a(r - r_e) + \frac{a^2}{2}(r - r_e)^2 + \cdots$$

and substitute this into u(r) to obtain

$$u(r) = D_e a^2 (r - r_e)^2 + \cdots$$

This same result can also be obtained by using a Taylor series about $r = r_e$.

$$u(r) = u(r_e) + \left[\frac{du}{dr}\right]_{r=r_e}(r - r_e) + \frac{1}{2}\left[\frac{d^2u}{dr^2}\right]_{r=r_e}(r - r_e)^2 + \cdots$$

From Figure 9-37 we see that $u(r_e) = 0$ and that

$$\left[\frac{du}{dr}\right]_{r=r_e} = 0$$

By comparing the harmonic oscillator potential energy

$$u(r) = \frac{1}{2}k(r - r_e)^2$$

with above, we find that

$$k = 2D_e a^2$$

and that

$$\nu = \frac{a}{2\pi}\left[\frac{2D_e}{\mu}\right]^{1/2}$$

$$= \frac{(1.963 \times 10^{10}\ \text{m}^{-1})}{2\pi} \times \left[\frac{(2)(456 \times 10^3\ \text{J}\cdot\text{mol}^{-1})}{((1/2)\ \text{amu})(1.6606 \times 10^{-27}\ \text{kg}\cdot\text{amu}^{-1}) \times (6.022 \times 10^{23}\ \text{mol}^{-1})}\right]^{1/2}$$

$$= 1.33 \times 10^{14}\ \text{s}^{-1} = 4.44 \times 10^3\ \text{cm}^{-1}$$

**9-10.**

$$I = \frac{1}{2}\iint d\sigma_1 d\sigma_2 [\alpha^*(1)\beta^*(2)\alpha(1)\beta(2) \pm \alpha^*(1)\beta^*(2)\alpha(2)\beta(1) \pm \alpha^*(2)\beta^*(1)\alpha(1)\beta(2) + \alpha^*(2)\beta^*(1)\alpha(2)\beta(1)]$$

$$= \frac{1}{2}\{1 \pm 0 \pm 0 + 1\} = 1$$

**9-11.** We must first evaluate the quantities $H_{ij}$ and $S_{ij}$.

$$H_{11} = \iint d\tau_1 d\tau_2 \psi_1^* \hat{H} \psi_1$$

$$= H_{11}^{\circ} \int d\sigma_1 \alpha^*(1)\alpha(1) \int d\sigma_2 \beta^*(2)\beta(2)$$

$$- H_{12}^{\circ} \int d\sigma_1 \beta^*(1)\alpha(1) \int d\sigma_2 \alpha^*(2)\beta(2)$$

$$- H_{12}^{\circ} \int d\sigma_1 \alpha^*(1)\beta(1) \int d\sigma_2 \beta^*(2)\alpha(2)$$

$$+ H_{11}^{\circ} \int d\sigma_1 \beta^*(1)\beta(1) \int d\sigma_2 \alpha^*(2)\alpha(2)$$

$$= 2H_{11}^{\circ} = -2 + 2J$$

where $H_{ij}^{\circ}$ are the quantities defined in Section 9-3. Similarly

$$H_{22} = 2H_{22}^{\circ} = -2 + 2J$$

$$H_{12} = H_{21} = -2H_{12}^{\circ} = 2S^2 - 2K$$

$$S_{11} = S_{22} = 2$$

$$S_{12} = -2S^2$$

The secular determinant is

$$\begin{vmatrix} 2H_{11}^{\circ} - 2E & -2H_{12}^{\circ} + 2ES^2 \\ -2H_{12}^{\circ} + 2ES^2 & 2H_{11}^{\circ} - 2E \end{vmatrix}$$

$$= \begin{vmatrix} H_{11}^{\circ} - E & -H_{12}^{\circ} + ES^2 \\ -H_{12}^{\circ} + ES^2 & H_{11}^{\circ} - E \end{vmatrix}$$

$$= \begin{vmatrix} -1 + J - E & S^2 - K + ES^2 \\ S^2 - K + ES^2 & -1 + J - E \end{vmatrix} = 0$$

The values of E are

$$E_{\pm} = -1 + \frac{J \pm K}{1 \pm S^2}$$

The algebraic equations that lead to this secular determinant are

$$c_1(-1 + J - E_{\pm}) + c_2(S^2 - K + E_{\pm}S^2) = 0$$

Using $E_+$, we have

$$c_1\left[\frac{JS^2 - K}{1 + S^2}\right] + c_2\left[\frac{-K + S^2J}{1 + S^2}\right] = 0$$

or

$$c_1 = - c_2 \qquad \text{(for } E_+\text{)}$$

If $E_-$ is substituted into the algebraic equations, we obtain

$$c_1 = c_2 \qquad \text{(for } E_-\text{)}$$

Thus we have that

$$\psi_+ = c_+(\psi_1 - \psi_2)$$

and

$$\psi_- = c_-(\psi_1 + \psi_2)$$

We can determine $c_+$ and $c_-$ by requiring that $\psi_+$ and $\psi_-$ be normalized.

$$\iint d\tau_1 d\tau_2 \psi_+^* \psi_+ = 1$$

$$= \iint d\tau_1 d\tau_2 [\psi_1^*\psi_1 - \psi_1^*\psi_2 - \psi_2^*\psi_1 + \psi_2^*\psi_2]$$

$$\begin{aligned}
&= c_+^2 \iint d\tau_1 d\tau_2 [1s_A\alpha^*(1)1s_B\beta^*(2) - 1s_A\alpha^*(2)1s_B\beta^*(1)] \\
&\qquad \times [1s_A\alpha(1)1s_B\beta(2) - 1s_A\alpha(2)1s_B\beta(1)] \\
&\quad - c_+^2 \iint d\tau_1 d\tau_2 [1s_A\alpha^*(1)1s_B\beta^*(2) - 1s_A\alpha^*(2)1s_B\beta^*(1)] \\
&\qquad \times [1s_A\beta(1)1s_B\alpha(2) - 1s_A\beta(2)1s_B\alpha(1)] \\
&\quad - c_+^2 \iint d\tau_1 d\tau_2 [1s_A\beta^*(1)1s_B\alpha^*(2) - 1s_A\beta^*(2)1s_B\alpha^*(1)] \\
&\qquad \times [1s_A\alpha(1)1s_B\beta(2) - 1s_A\alpha(2)1s_B\beta(1)] \\
&\quad + c_+^2 \iint d\tau_1 d\tau_2 [1s_A\beta^*(1)1s_B\alpha^*(2) - 1s_A\beta^*(2)1s_B\alpha^*(1)] \\
&\qquad \times [1s_A\beta(1)1s_B\alpha(2) - 1s_A\beta(2)1s_B\alpha(1)]
\end{aligned}$$

$$= c_+^2(2 + 2S^2 + 2S^2 + 2)$$

$$= c_+^2(4 + 4S^2) = 1$$

or

$$c_+ = \frac{1}{\sqrt{4 + 4S^2}} = \frac{1}{\sqrt{2 + 2S^2}} \times \frac{1}{\sqrt{2}}$$

and so

$$\psi_+ = \frac{1}{\sqrt{2 + 2S^2}}\left[\frac{1}{\sqrt{2}}\psi_1 - \frac{1}{\sqrt{2}}\psi_2\right]$$

Similarly,

$$c_-^2(4 - 4S^2) = 1$$

and

$$\psi_- = \frac{1}{\sqrt{2 - 2S^2}}\left[\frac{1}{\sqrt{2}}\psi_1 + \frac{1}{\sqrt{2}}\psi_2\right]$$

Lastly, we calculate the energy associated with $\psi_3$ and $\psi_4$.

$$E_3 = \iint d\tau_1 d\tau_2 \psi_3^* \hat{H} \psi_3$$

$$= \frac{1}{2 - 2S^2}\iint d\tau_1 d\tau_2 [1s_A\alpha^*(1)1s_B\alpha^*(2) - 1s_A\alpha^*(2)1s_B\alpha^*(1)]$$
$$\times \hat{H}[1s_A\alpha(1)1s_B\alpha(2) - 1s_A\alpha(2)1s_B\alpha(1)]$$

$$= \frac{1}{2 - 2S^2}(H_{11}^{o} - H_{12}^{o} - H_{12}^{o} + H_{11}^{o})$$

$$= \frac{H_{11}^{o} - H_{12}^{o}}{1 - S^2}$$

$$= \frac{-1 + J + S^2 - K}{1 - S^2} = -1 + \frac{J - K}{1 - S^2} = E_-$$

The proof that $E_4 = E_-$ is similar.

**9-12.**

$$\psi_S = \frac{1}{\sqrt{2 + 2S^2}}\left[\frac{1}{\sqrt{2}}\psi_1 - \frac{1}{\sqrt{2}}\psi_2\right]$$

We must evaluate $\hat{S}_z\psi_1$ and $\hat{S}_z\psi_2$.

$$\hat{S}_z\psi_1 = (\hat{S}_{z1} + \hat{S}_{z2})[1s_A\alpha(1)1s_B\beta(2) - 1s_A\alpha(2)1s_B\beta(1)]$$

$$= \left[+\frac{1}{2}\hbar - \frac{1}{2}\hbar\right]1s_A\alpha(1)1s_B\beta(2) - \left[-\frac{1}{2}\hbar + \frac{1}{2}\hbar\right]1s_A\alpha(2)1s_B\beta(1) = 0$$

$$\hat{S}_z\psi_2 = (\hat{S}_{z1} + \hat{S}_{z2})[1s_A\beta(1)1s_B\alpha(2) - 1s_A\beta(2)1s_B\alpha(1)]$$

$$= \left[-\frac{1}{2}\hbar + \frac{1}{2}\hbar\right]1s_A\beta(1)1s_B\alpha(2) - \left[+\frac{1}{2}\hbar - \frac{1}{2}\hbar\right]1s_A\beta(2)1s_B\alpha(1) = 0$$

Thus

$$\hat{S}_z\psi_S = 0$$

Since

$$\psi_t = \frac{1}{\sqrt{2(1 - S^2)}}\left[\frac{1}{\sqrt{2}}\psi_1 + \frac{1}{\sqrt{2}}\psi_2\right]$$

we also have $\hat{S}_z\psi_t = 0$.

Now let's look at $\hat{S}_z$ acting upon Eqs. 9-49 and 9-50.

Consider

$$\hat{S}_z\begin{vmatrix} 1s_A\alpha(1) & 1s_B\alpha(1) \\ 1s_A\alpha(2) & 1s_B\alpha(2) \end{vmatrix}$$

$$= (\hat{S}_{z1} + \hat{S}_{z2})[1s_A\alpha(1)1s_B\alpha(2) - 1s_A\alpha(2)1s_B\alpha(1)]$$

$$= \left[+\frac{1}{2}\hbar + \frac{1}{2}\hbar\right]1s_A\alpha(1)1s_B\alpha(2) - \left[+\frac{1}{2}\hbar + \frac{1}{2}\hbar\right]1s_A\alpha(2)1s_B\alpha(1)$$

$$= +\hbar\begin{vmatrix} 1s_A\alpha(1) & 1s_B\alpha(1) \\ 1s_A\alpha(2) & 1s_B\alpha(2) \end{vmatrix}$$

Similarly

$$\hat{S}_z\begin{vmatrix} 1s_A\beta(1) & 1s_B\beta(1) \\ 1s_A\beta(2) & 1s_B\beta(2) \end{vmatrix}$$

$$= \left[-\frac{1}{2}\hbar - \frac{1}{2}\hbar\right]1s_A\beta(1)1s_B\beta(2) - \left[-\frac{1}{2}\hbar - \frac{1}{2}\hbar\right]1s_A\beta(2)1s_B\beta(1)$$

$$= -\hbar\begin{vmatrix} 1s_A\beta(1) & 1s_B\beta(1) \\ 1s_A\beta(2) & 1s_B\beta(2) \end{vmatrix}$$

**9-13.** The Hamiltonian operator of $H_2^+$ is

$$\hat{H} = -\frac{1}{2}\nabla^2 - \frac{1}{r_A} - \frac{1}{r_B} + \frac{1}{R}$$

Thus

$$H_{AA} = \int d\mathbf{r}\, 1s_A \hat{H} 1s_A$$

$$= \int d\mathbf{r}\, 1s_A \left[-\frac{1}{2}\nabla^2 - \frac{1}{r_A} - \frac{1}{r_B} + \frac{1}{R}\right] 1s_A$$

$$= \int d\mathbf{r}\, 1s_A \left[-\frac{1}{2} - \frac{1}{r_B} + \frac{1}{R}\right] 1s_A$$

$$= -\frac{1}{2} + J'$$

$$H_{AB} = \int d\mathbf{r}\, 1s_A \left[-\frac{1}{2}\nabla^2 - \frac{1}{r_A} - \frac{1}{r_B} + \frac{1}{R}\right] 1s_B$$

$$= \int d\mathbf{r}_2\, 1s_A \left[-\frac{1}{2} - \frac{1}{r_A} + \frac{1}{R}\right] 1s_B$$

$$= -\frac{S}{2} + K'$$

**9-14.** The overlap integral is derived in Problem 9-3.

$$J' = \int d\mathbf{r}\, 1s_A \left[-\frac{1}{r_B} + \frac{1}{R}\right] 1s_A$$

$$= \frac{R^3}{8\pi}\int_0^{2\pi} d\phi \int_1^{\infty} d\lambda \int_{-1}^{1} d\mu (\lambda^2 - \mu^2) e^{-R(\lambda+\mu)} \left[-\frac{1}{R(\lambda - \mu)/2}\right] + \frac{1}{R}$$

$$= -\frac{R^2}{2}\int_1^{\infty} d\lambda e^{-R\lambda} \int_{-1}^{1} d\mu (\lambda + \mu) e^{-R\mu} + \frac{1}{R}$$

$$= -\frac{R^2}{2}\int_1^{\infty} d\lambda e^{-R\lambda}\left[\lambda\left[\frac{e^R - e^{-R}}{R}\right] + \frac{e^R(1 - R) - e^{-R}(R + 1)}{R^2}\right] + \frac{1}{R}$$

$$= -\frac{R^2}{2}\left[\left[\frac{e^R - e^{-R}}{R}\right]\left[\frac{R + 1}{R^2}\right]e^{-R} + \left[\frac{e^R(1 - R) - e^{-R}(R + 1)}{R^2}\right]\frac{e^{-R}}{R}\right] + \frac{1}{R}$$

$$= -\frac{1}{2R}\left[2 - 2e^{-2R}(1 + R)\right] + \frac{1}{R}$$

$$= -\frac{1}{R} + e^{-2R}\left[1 + \frac{1}{R}\right] + \frac{1}{R}$$

$$= e^{-2R}\left[1 + \frac{1}{R}\right]$$

$$K' = \int d\mathbf{r}\, 1s_A\left[-\frac{1}{r_A} + \frac{1}{R}\right]1s_B$$

$$= -\frac{R^3}{4}\int_1^{\infty} d\lambda \int_{-1}^{1} d\mu(\lambda^2 - \mu^2)\frac{e^{-R\lambda}}{R(\lambda - \mu)/2} + \frac{S}{R}$$

$$= -\frac{R^2}{2}\int_1^{\infty} d\lambda e^{-R\lambda}\int_{-1}^{1} d\mu(\lambda + \mu) + \frac{S}{R}$$

$$= -R^2\int_1^{\infty} d\lambda\ \lambda e^{-R\lambda} + \frac{S}{R}$$

$$= -R^2\left[\frac{e^{-R}(R + 1)}{R^2}\right] + \frac{S}{R} = \frac{S}{R} - e^{-R}(R + 1)$$

**9-15.** Multiply out $\psi_{MO}$ to obtain

$$\psi_{MO} = c_b\left[1s_A(1)1s_A(2) + 1s_A(1)1s_B(2) + 1s_B(1)1s_A(2) + 1s_B(1)1s_B(2)\right]$$
$$+ c_a\left[1s_A(1)1s_A(2) - 1s_A(1)1s_B(2) - 1s_B(1)1s_A(2) + 1s_B(1)1s_B(2)\right]$$

$$= (c_b - c_a)\left[1s_A(1)1s_A(2) + 1s_A(2)1s_B(1)\right]$$
$$+ (c_b + c_a)\left[1s_A(1)1s_A(2) + 1s_B(1)1s_B(2)\right]$$

By comparing $\psi_{VB}$ and $\psi_{MO}$, we see that

$$c_n = c_b - c_a \qquad\qquad c_i = c_b + c_a$$

Therefore,

$$\frac{c_i}{c_n} = \frac{c_b + c_a}{c_b - c_a} = \frac{1 + (c_a/c_b)}{1 - (c_a/c_b)}$$

If $c_a/c_b = -0.73$, then $c_i/c_n = 0.16$.

**9-16.** The argument that $H_{12} = 0$ is given on p. 367. Now consider $S_{12}$.

$$S_{12} = \iint dr_1 dr_2 \psi_1 \psi_2$$

$$= \iint dr_1 dr_2 \phi_b(1)\phi_b(2)[\phi_b(1)\phi_a(2) + \phi_a(1)\phi_b(2)]$$

Both terms here contain a factor of the form

$$\int dr_i \, \phi_b(i)\phi_a(i)$$

where i = 2 in the first term and i = 1 in the second term. Now

$$I = \int dr_i \, \phi_b(i)\phi_a(i) = \int dr_i \left[1s_A(i) + 1s_B(i)\right] \times \left[1s_A(i) - 1s_B(i)\right]$$

The letters A and B label the two identical nuclei in $H_2$, and so the value of I must remain the same under such an interchange. If we interchange the labels A and B, then we obtain

$$I = -I$$

which implies that I = 0.

**9-17.** The energy will be the same as that calculated from $\psi_1$ and $\psi_3$ separately if $H_{13} = 0$, since the two roots of

$$\begin{vmatrix} H_{11} - E & H_{13} \\ H_{13} & H_{33} - E \end{vmatrix} = 0$$

will be simply $H_{11}$ and $H_{33}$ if $H_{13} = 0$. Thus we must prove that $H_{13} = 0$.

$$H_{13} = \iint dr_1 dr_2 \psi_1 \hat{H} \psi_3$$

$$= \iint dr_1 dr_2 \phi_b(1)\phi_b(2)\hat{H}[\phi_b(1)\phi_a(2) - \phi_a(1)\phi_b(2)]$$

The two integrals here contain factors of the form

$$I = \int dr_i\, \phi_b(i)\hat{H}\phi_a(i)$$

but this integral is equal to zero by the same argument as that given on p. 367 for Eq. 9-88 (see also Problem 9-16).

**9-18.** If

$$\psi = c_1\chi_A + c_2\chi_B$$

then the secular determinantal equation is

$$\begin{vmatrix} \alpha_A - E & \beta - ES \\ \beta - ES & \alpha_B - E \end{vmatrix} = 0$$

and

$$(\alpha_A - E)(\alpha_B - E) - (\beta - ES)^2 = 0$$

or

$$(1 - S^2)E^2 + [2\beta S - \alpha_A - \alpha_B]E + \alpha_A\alpha_B - \beta^2 = 0$$

The two roots neglecting S are

$$E = \frac{\alpha_A + \alpha_B \pm \left[(\alpha_A + \alpha_B)^2 - 4\alpha_A\alpha_B + 4\beta^2\right]^{1/2}}{2}$$

$$= \frac{\alpha_A + \alpha_B \pm \left[(\alpha_A - \alpha_B)^2 + 4\beta^2\right]^{1/2}}{2}$$

If $\chi_A$ and $\chi_B$ correspond to the same energy, then $\alpha_A = \alpha_B$ and

$$E = \alpha \pm \beta$$

If $\alpha_A \neq \alpha_B$, then (assuming that $\alpha_A < \alpha_B$)

$$E_{\pm} = \frac{\alpha_A + \alpha_B}{2} \pm \frac{(\alpha_A - \alpha_B)}{2}\left[1 + \frac{4\beta^2}{(\alpha_A - \alpha_B)^2}\right]^{1/2}$$

$$= \frac{\alpha_A + \alpha_B}{2} \pm \frac{(\alpha_A - \alpha_B)}{2} \pm \frac{\beta^2}{(\alpha_A - \alpha_B)} + \cdots$$

Therefore

$$E_+ = \alpha_A + \frac{\beta^2}{\alpha_A - \alpha_B} + \cdots$$

and

$$E_- = \alpha_B - \frac{\beta^2}{\alpha_A - \alpha_B}$$

Note that if the difference between $\alpha_A$ and $\alpha_B$ is large, then the two energies obtained are simply $\alpha_A$ and $\alpha_B$. There is no lowering of the energy.

**9-19.** The electron configuration of $N_2$ and $N_2^+$ are

$N_2$: $KK(\sigma 2s)^2(\sigma^*2s)^2(\pi 2p)^4(\sigma 2p_z)^2$

$N_2^+$: $KK(\sigma 2s)^2(\sigma^*2s)^2(\pi 2p)^4(\sigma 2p_z)^1$

The bond order of $N_2$ is 3 and that of $N_2^+$ is 2 1/2. Thus the dissociation energy of $N_2$ is greater than that of $N_2^+$.

The electron configurations of $O_2$ and $O_2^+$ are

$O_2$: $KK(\sigma 2s)^2(\sigma^*2s)^2(\pi 2p)^4(\sigma 2p_z)^2(\pi^*2p)^2$

$O_2^+$: $KK(\sigma 2s)^2(\sigma^*2s)^2(\pi 2p)^4(\sigma 2p_z)^2(\pi^*2p)^1$

The species $O_2^+$ has one less antibonding electron than $O_2$ and so the dissociation energy of $O_2^+$ is greater than that of $O_2$.

**9-20.** The electron configurations of $F_2$ and $F_2^+$ are

$F_2$: $KK(\sigma 2s)^2(\sigma^*2s)^2(\pi 2p)^4(\sigma 2p_z)^2(\pi^*2p)^4$

$F_2^+$: $KK(\sigma 2s)^2(\sigma^*2s)^2(\pi 2p)^4(\sigma 2p_z)^2(\pi^*2p)^3$

$F_2^+$ has one less antibonding electron than $F_2$ and so $F_2^+$ has a larger bond strength and a shorter bond than $F_2$.

**9-21.** The electron configurations of $N_2^+$, $N_2$ and $N_2^-$ are

$N_2^+$: $KK(\sigma 2s)^2(\sigma^*2s)^2(\pi 2p)^4(\sigma 2p_z)^1$

$N_2$: $KK(\sigma 2s)^2(\sigma^*2s)^2(\pi 2p)^4(\sigma 2p_z)^2$

$N_2^-$: $KK(\sigma 2s)^2(\sigma^*2s)^2(\pi 2p)^4(\sigma 2p_z)^2(\pi^*2p)^1$

The bond orders of $N_2^+$, $N_2$ and $N_2^-$ are 2 1/2, 3 and 2 1/2, respectively. Thus we predict that the relative stabilities should go as

$N_2 > N_2^+ \approx N_2^-$

**9-22.** The electron configurations of $C_2$ and $C_2^{2-}$ are

$C_2$: $KK(\sigma 2s)^2(\sigma^* 2s)^2(\pi 2p)^4$

$C_2^{2-}$: $KK(\sigma 2s)^2(\sigma^* 2s)^2(\pi 2p)^4(\sigma 2p_z)^2$

The acetylide ion $C_2^{2-}$ has two more bonding electrons than $C_2$ and so we predict that $C_2^{2-}$ has a shorter bond length and a stronger bond than $C_2$.

**9-23.**
$Na_2$ $KKLL(\sigma 3s)^2$

$Mg_2$ $KKLL(\sigma 3s)^2(\sigma^* 3s)^2$ (not stable)

$Al_2$ $KKLL(\sigma 3s)^2(\sigma^* 3s)^2(\pi 3p)^2$

$Si_2$ $KKLL(\sigma 3s)^2(\sigma^* 3s)^2(\pi 3p)^4$

$P_2$ $KKLL(\sigma 3s)^2(\sigma^* 3s)^2(\pi 3p)^4(\sigma 3p_z)^2$

$S_2$ $KKLL(\sigma 3s)^2(\sigma^* 3s)^2(\pi 3p)^4(\sigma 3p_z)^2(\pi^* 3p)^2$

$Cl_2$ $KKLL(\sigma 3s)^2(\sigma^* 3s)^2(\pi 3p)^4(\sigma 3p_z)^2(\pi^* 3p)^4$

$Ar_2$ $KKLL(\sigma 3s)^2(\sigma^* 3s)^2(\pi 3p)^4$
$(\sigma 3p_z)^2(\pi^* 3p)^4(\sigma^* 3p_z)^2$ (not stable)

**9-24.**
$NO^+$ $KK(\sigma 2s)^2(\sigma^* 2s)^2(\pi 2p)^4(\sigma 2p_z)^2$

NO $KK(\sigma 2s)^2(\sigma^* 2s)^2(\pi 2p)^4(\sigma 2p_z)^2(\pi^* 2p)^1$

The bond orders of $NO^+$ and NO are 3 and 2 1/2, respectively.

**9-25.** The electron configuration of $CN^-$ is

$CN^-$: $KK(\sigma 2s)^2(\sigma^* 2s)^2(\pi 2p)^4(\sigma 2p_z)^2$

The bond order of $CN^-$ is 3.

**9-26.** A filled $\sigma$ or $\sigma^*$ orbital has

$$M_L = 0 + 0 = 0$$

and

$$M_S = + 1/2 - 1/2 = 0.$$

Filled $\pi$ orbitals (doubly degenerate) have

$$M_L = + 1 - 1 + 1 - 1 = 0$$

and

$$M_S = + 1/2 - 1/2 + 1/2 - 1/2 = 0.$$

**9-27.**

| | | |
|---|---|---|
| $H_2^+$: | $M_L = 0$, $M_S = \pm 1/2$ | $^2\Sigma$ |
| $H_2$: | $M_L = 0 + 0$, $M_S = + 1/2 - 1/2$ | $^1\Sigma$ |
| $He_2^+$: | $M_L = 0 + 0 + 0$, $M_S = \pm 1/2$ | $^2\Sigma$ |
| $Li_2$: | $M_L = 0$, $M_S = 0$ | $^1\Sigma$ |
| $B_2$: | see p. 388 | $^3\Sigma$ |
| $C_2$: | all doubly occupied orbitals | $^1\Sigma$ |
| $N_2^+$: | $M_L = 0$, $M_S = \pm 1/2$ | $^2\Sigma$ |
| $N_2$: | all doubly occupied orbitals | $^1\Sigma$ |
| $O_2^+$: | $M_L = \pm 1$, $M_S = \pm 1/2$ | $^2\Pi$ |
| $O_2$: | There are two electrons beyond the completely filled shells. The situation is similar to that of $B_2$. | |
| $F_2$: | all doubly occupied orbitals | $^1\Sigma$ |

**9-28.** See Problem 9-27.

**9-29.** Set up the following table.

| | $m_{\ell_1}$ | $m_{s_1}$ | $m_{\ell_2}$ | $m_{s_2}$ | $M_L$ | $M_S$ |
|---|---|---|---|---|---|---|
| 1. | + 1 | + 1/2 | 0 | + 1/2 | 1 | 1 |
| 2. | + 1 | + 1/2 | 0 | − 1/2 | 1 | 0 |
| 3. | + 1 | − 1/2 | 0 | + 1/2 | 1 | 0 |
| 4. | + 1 | − 1/2 | 0 | − 1/2 | 1 | − 1 |
| 5. | − 1 | + 1/2 | 0 | + 1/2 | − 1 | 1 |
| 6. | − 1 | + 1/2 | 0 | − 1/2 | − 1 | 0 |
| 7. | − 1 | − 1/2 | 0 | + 1/2 | − 1 | 0 |
| 8. | − 1 | − 1/2 | 0 | − 1/2 | − 1 | − 1 |

Entries 1, 2, 4, 5, 6, and 8 correspond to $|M_L| = 1$ and $S = 1$, giving a $^3\Pi$ state. Entries 3 and 7 correspond to $|M_L| = 1$ and $S = 0$, giving a $^1\Pi$ state.

**9-30.** A covalent valence-bond wave function for HF, including the 1s, 2s and 2p orbitals on fluorine is

$$\begin{aligned}\psi_{cov} = &\left|\psi_{1sF}\alpha(1)\psi_{1sF}\beta(2)\psi_{2sF}\alpha(3)\psi_{2sF}\beta(4)\right.\\ &\psi_{2p_xF}\alpha(5)\psi_{2p_xF}\beta(6)\psi_{2p_yF}\alpha(7)\\ &\left.\psi_{2p_yF}\beta(8)\psi_{2p_zF}\alpha(9)\psi_{1sH}\beta(10)\right|\\ - &\left|\psi_{1sF}\alpha(1)\psi_{1sF}\beta(2)\psi_{2sF}\alpha(3)\psi_{2sF}\beta(4)\right.\\ &\psi_{2p_xF}\alpha(5)\psi_{2p_xF}\beta(6)\psi_{2p_yF}\alpha(7)\\ &\left.\psi_{2p_yF}\beta(8)\psi_{2p_zF}\beta(9)\psi_{1sH}\alpha(10)\right|\end{aligned}$$

**9-31.** By analogy with Eq. 9-112, a covalent valence-bond wave function for $NH_3$ is

$$\begin{aligned}\psi = &\left|\psi_{2p_xN}\alpha(1)\psi_{1sA}\beta(2)\psi_{2p_yN}\alpha(3)\psi_{1sB}\beta(4)\psi_{2p_zN}\alpha(5)\psi_{1sC}\beta(6)\right|\\ - &\left|\psi_{2p_xN}\beta(1)\psi_{1sA}\alpha(2)\psi_{2p_yN}\alpha(3)\psi_{1sB}\beta(4)\psi_{2p_zN}\alpha(5)\psi_{1sC}\beta(6)\right|\\ + &\left|\psi_{2p_xN}\beta(1)\psi_{1sA}\alpha(2)\psi_{2p_yN}\beta(3)\psi_{1sB}\alpha(4)\psi_{2p_zN}\alpha(5)\psi_{1sC}\beta(6)\right|\\ - &\left|\psi_{2p_xN}\beta(1)\psi_{1sA}\alpha(2)\psi_{2p_yN}\beta(3)\psi_{1sB}\alpha(4)\psi_{2p_zN}\beta(5)\psi_{1sC}\alpha(6)\right|\\ + &\left|\psi_{2p_xN}\alpha(1)\psi_{1sA}\beta(2)\psi_{2p_yN}\beta(3)\psi_{1sB}\alpha(4)\psi_{2p_zN}\beta(5)\psi_{1sC}\alpha(6)\right|\\ - &\left|\psi_{2p_xN}\alpha(1)\psi_{1sA}\beta(2)\psi_{2p_yN}\alpha(3)\psi_{1sB}\beta(4)\psi_{2p_zN}\beta(5)\psi_{1sC}\alpha(6)\right|\\ + &\left|\psi_{2p_xN}\beta(1)\psi_{1sA}\alpha(2)\psi_{2p_yN}\alpha(3)\psi_{1sB}\beta(4)\psi_{2p_zN}\beta(5)\psi_{1sC}\alpha(6)\right|\\ - &\left|\psi_{2p_xN}\alpha(1)\psi_{1sA}\beta(2)\psi_{2p_yN}\beta(3)\psi_{1sB}\alpha(4)\psi_{2p_zN}\alpha(5)\psi_{1sC}\beta(6)\right|\end{aligned}$$

**9-32.** We start with

$$\xi_1 = \frac{1}{\sqrt{3}}2s + \sqrt{\frac{2}{3}}\,2p_z$$

Requiring

$$\xi_2 = \frac{1}{\sqrt{3}}2s + b_2 2p_z + c_2 2p_x$$

to be orthogonal to $\xi_1$ yields

$$\frac{1}{3} + b_2\sqrt{\frac{2}{3}} = 0$$

or

$$b_2 = -\frac{1}{\sqrt{6}}$$

Requiring that $\xi_2$ be normalized yields

$$\frac{1}{3} + \frac{1}{6} + c_2^2 = 1$$

or

$$c_2 = \frac{1}{\sqrt{2}}$$

The resulting hybrid orbital is

$$\xi_2 = \frac{1}{\sqrt{3}}2s - \frac{1}{\sqrt{6}}2p_z + \frac{1}{\sqrt{2}}2p_x$$

**9-33.** We start with

$$\xi_1 = \frac{1}{\sqrt{3}}2s + \sqrt{\frac{2}{3}}\,2p_z$$

and

$$\xi_2 = \frac{1}{\sqrt{3}}2s - \frac{1}{\sqrt{6}}2p_z + \frac{1}{\sqrt{2}}2p_x$$

Now write

$$\xi_3 = a_3 2s + b_3 2p_z + c_3 2p_x$$

Requiring that $\xi_3$ be orthogonal to $\xi_1$ and $\xi_2$ yields

$$\frac{a_3}{\sqrt{3}} + b_3\sqrt{\frac{2}{3}} = 0$$

$$\frac{a_3}{\sqrt{3}} - \frac{b_3}{\sqrt{6}} + \frac{c_3}{\sqrt{2}} = 0$$

and requiring that $\xi_3$ be normalized yields

$$a_3^2 + b_3^2 + c_3^2 = 1$$

The first of these three requirements gives

$$a_3 = -b_3\sqrt{2}$$

Substituting this result into the second requirement gives

$$c_3 = \sqrt{3}b_3$$

and the normalization condition yields

$$b_3^2(2 + 1 + 3) = 1$$

or

$$b_3 = \pm\frac{1}{\sqrt{6}}$$

Thus (taking the negative root so that $a_3$ is positive)

$$\xi_3 = \frac{1}{\sqrt{3}}2s - \frac{1}{\sqrt{6}}2p_z - \frac{1}{\sqrt{2}}2p_x$$

**9-34.** Start with

$$\tan\theta = -\sqrt{8}$$

Thus

$$\frac{\sin^2\theta}{\cos^2\theta} = \frac{1 - \cos^2\theta}{\cos^2\theta} = 8$$

or

$$\cos^2\theta = 1/9$$

or

$$\cos\theta = \pm 1/3$$

We take the negative root because $\theta$ is in the second quadrant.

**9-35.** Because $\xi_1$ points along the z axis, we may take the second orbital to lie in any plane containing the z axis. Let this plane be the x-z plane and write

$$\xi_2 = a_2 2s + b_2 2p_x + d_2 2p_z$$

which lies in the x-z plane.

If we let $\phi = 0$ in Eq. 9-123, then we obtain

$$\psi_{2p_x} = \left[\frac{3}{4\pi}\right]^{1/2} R(r)\sin\theta$$

Requiring that $\xi_1$ and $\xi_2$ be orthogonal yields

$$\frac{a_2}{2} + \frac{d_2\sqrt{3}}{2} = 0$$

or simply that $a_2 = - d_2\sqrt{3}$. If we require $\xi_2$ to be normalized, then we have

$$a_2^2 + b_2^2 + d_2^2 = 1$$

or

$$3d_2^2 + b_2^2 + d_2^2 = 1$$

or

$$b_2 = (1 - 4d_2^2)^{1/2}$$

Thus $\xi_2$ becomes

$$\xi_2 = \left[\frac{3}{4\pi}\right]^{1/2} R(r)[- d_2 + (1 - 4d_2^2)^{1/2}\sin\theta + d_2\cos\theta]$$

Maximizing $\xi_2$ with respect to $\theta$ gives

$$\tan\theta = \frac{(1 - 4d_2^2)^{1/2}}{d_2}$$

and maximizing with respect to $d_2$ yields

$$- 1 - \frac{4d_2}{(1 - 4d_2^2)^{1/2}}\sin\theta + \cos\theta = 0$$

The first of these two relations gives

$$\frac{d_2}{(1 - 4d_2^2)^{1/2}} = \frac{\cos\theta}{\sin\theta}$$

Substitute this into the above equation to get

$$- 1 - 4\cos\theta + \cos\theta = 0$$

or

$$\cos\theta = - 1/3$$

**9-36.** We expand the determinant about its first row according to (expansion in minors)

$$\begin{vmatrix} x & 1 & 0 & 0 \\ 1 & x & 1 & 0 \\ 0 & 1 & x & 1 \\ 0 & 0 & 1 & x \end{vmatrix} =$$

$$= x\begin{vmatrix} x & 1 & 0 \\ 1 & x & 1 \\ 0 & 1 & x \end{vmatrix} - \begin{vmatrix} 1 & 1 & 0 \\ 0 & x & 1 \\ 0 & 1 & x \end{vmatrix}$$

$$= x(x^3 - 2x) - (x^2 - 1) = x^4 - 3x^2 + 1$$

**9-37.** Let cyclobutadiene be represented by

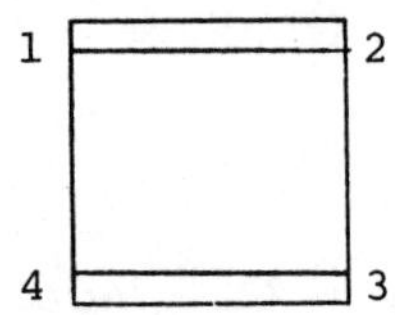

The Hückel determinantal equation is

$$\begin{vmatrix} x & 1 & 0 & 1 \\ 1 & x & 1 & 0 \\ 0 & 1 & x & 1 \\ 1 & 0 & 1 & x \end{vmatrix}$$

$$= x\begin{vmatrix} x & 1 & 0 \\ 1 & x & 1 \\ 0 & 1 & x \end{vmatrix} - \begin{vmatrix} 1 & 1 & 0 \\ 0 & x & 1 \\ 1 & 1 & x \end{vmatrix}$$

$$- \begin{vmatrix} 1 & x & 1 \\ 0 & 1 & x \\ 1 & 0 & 1 \end{vmatrix}$$

$$= x(x^3 - 2x) - (x^2 + 1 - 1) - (1 + x^2 - 1)$$
$$= x^4 - 4x^2 = 0$$

The four roots to this equation are $x = 2, 0, 0, -2$ and the four energies are ($x = (\alpha - E)/\beta$)

$$E = \begin{cases} \alpha - 2\beta \\ \alpha \\ \alpha \\ \alpha + 2\beta \end{cases}$$

The π-electron energy is (there are four π electrons)

$$E_\pi = 2(\alpha + 2\beta) + 2\alpha = 4\alpha + 4\beta$$

Because of Hund's rule, the ground state is predicted to be a triplet. The delocalization energy of cyclobutadiene is

$$DE = E_\pi(\text{cyclobutadiene}) - 2E_\pi(\text{ethylene}) = 0$$

**9-38.** The Hückel determinantal equation of trimethylenemethane is (atom 4 is the central carbon atom)

$$\begin{vmatrix} x & 0 & 0 & 1 \\ 0 & x & 0 & 1 \\ 0 & 0 & x & 1 \\ 1 & 1 & 1 & x \end{vmatrix}$$

$$= x \begin{vmatrix} x & 0 & 1 \\ 0 & x & 1 \\ 1 & 1 & x \end{vmatrix} - \begin{vmatrix} 0 & x & 0 \\ 0 & 0 & x \\ 1 & 1 & 1 \end{vmatrix}$$

$$= x(x^3 - 2x) - (x^2) = x^4 - 3x^2 = 0$$

The four roots are $x = \sqrt{3}, 0, 0, -\sqrt{3}$ and the four energies are

$$E = \begin{cases} \alpha - \sqrt{3}\beta \\ \alpha \\ \alpha \\ \alpha + \sqrt{3}\beta \end{cases}$$

There are four π electrons and so the π-electron energy is

$$E_\pi = 2(\alpha + \sqrt{3}\beta) + 2\alpha = 4\alpha + 2\sqrt{3}\beta$$

The energy of two isolated ethylene molecules is $4\alpha + 4\beta$ and so

$$DE = E_\pi(\text{trimethylenemethane}) - 2E_\pi(\text{ethylene})$$
$$= -0.536\ \beta$$

**9-39.** Number the carbon atoms according to

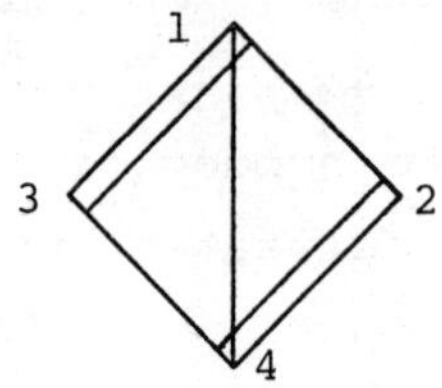


The Hückel determinantal equation is

$$\begin{vmatrix} x & 1 & 1 & 1 \\ 1 & x & 0 & 1 \\ 1 & 0 & x & 1 \\ 1 & 1 & 1 & x \end{vmatrix} = 0$$

If we expand this determinant in minors about the second row, then we obtain

$$-\begin{vmatrix} 1 & 1 & 1 \\ 0 & x & 1 \\ 1 & 1 & x \end{vmatrix} + x\begin{vmatrix} x & 1 & 1 \\ 1 & x & 1 \\ 1 & 1 & x \end{vmatrix} + \begin{vmatrix} x & 1 & 1 \\ 1 & 0 & x \\ 1 & 1 & 1 \end{vmatrix}$$

$$= -(x^2 + 1 - x - 1) + x(x^3 + 2 - 3x)$$
$$+ (x + 1 - 1 - x^2)$$
$$= x^4 - 5x^2 + 4x = 0$$

An obvious root to this equation is $x = 0$, leaving

$$x^3 - 5x + 4 = 0$$

We find the root $x = 1$ by inspection. Divide this cubic equation by $x - 1$ to obtain

$$
\begin{array}{r}
x^2 + x - 4 \\
x - 1 \overline{)\, x^3 - 5x + 4} \\
\underline{x^3 - x^2} \\
x^2 - 5x \\
\underline{x^2 - x} \\
-4x + 4
\end{array}
$$

Thus the above cubic equation factors as

$$(x - 1)(x^2 + x - 4) = 0$$

The two roots of the quadratic equation are

$$x = \frac{-1 \pm \sqrt{17}}{2}$$

The four roots of the Hückel determinantal equation are $x = 1, 0, -1/2 \pm (1/2)\sqrt{17}$.

The four energies are

$$E = \begin{cases} \alpha - 1.562\ \beta \\ \alpha - \beta \\ \alpha \\ \alpha + 2.562\ \beta \end{cases}$$

The $\pi$-electron energy is

$$E_\pi = 2(\alpha + 2.562\beta) + 2\alpha = 4\alpha + 5.124\ \beta$$

and the delocalization energy is

$$DE = E_\pi(\text{bicyclobutadiene}) - 2E_\pi(\text{ethylene})$$
$$= 1.124\ \beta$$

**9-40.** The Hückel determinantal equation is

$$\begin{vmatrix} x & 1 & 0 \\ 1 & x & 1 \\ 0 & 1 & x \end{vmatrix} = x^3 - 2x = 0$$

The three roots are $x = \sqrt{2}, 0, -\sqrt{2}$ and the three energies are

$$E = \begin{cases} \alpha - \sqrt{2}\beta \\ \alpha \\ \alpha + \sqrt{2}\beta \end{cases}$$

In order to calculate the charges on each carbon atom and the bond orders, we must determine the molecular orbital associated with each value of E. To do this, we must go back to the algebraic equations for $c_1$, $c_2$ and $c_3$ in

$$\psi = c_1\chi_1 + c_2\chi_2 + c_3\chi_3$$

that lead to the determinantal equation.
These equations are

$$c_1(\alpha - E) + c_2\beta = 0$$

$$c_1\beta + c_2(\alpha - E) + c_3\beta = 0$$

$$c_2\beta + c_3(\alpha - E) = 0$$

If we let $E = \alpha$, then we have

$$c_2 = 0$$

$$c_1 + c_3 = 0$$

$$c_2 = 0$$

from which we find that

$$\psi_2 = c_1(\chi_1 - \chi_3)$$

where we have used a subscript 2 because $E = \alpha$ is the second energy level.
Requiring that $\psi_2$ be normalized gives

$$c_1^2(1 + 1) = 1$$

or

$$c_1 = 1/\sqrt{2}$$

Thus we have

$$\psi_2 = (1/\sqrt{2})(\chi_1 - \chi_3)$$

Now let $E = \alpha + \sqrt{2}\,\beta$ and find

$$-\sqrt{2}c_1 + c_2 = 0$$

$$c_1 - \sqrt{2}c_2 + c_3 = 0$$

$$c_2 - \sqrt{2}c_3 = 0$$

from which we obtain

$$c_1 = c_3 = (1/\sqrt{2})c_2$$

Thus we have

$$\psi_1 = c_2\left[\frac{1}{\sqrt{2}}\chi_1 + \chi_2 + \frac{1}{\sqrt{2}}\chi_3\right]$$

Normalization gives

$$c_2^2\left[\frac{1}{2} + 1 + \frac{1}{2}\right] = 1$$

or

$$c_2 = 1/\sqrt{2}$$

Thus

$$\psi_1 = \frac{1}{2}\chi_1 + \frac{1}{\sqrt{2}}\chi_2 + \frac{1}{2}\chi_3$$

Lastly, let $E = \alpha - \sqrt{2}\ \beta$ to obtain

$$\sqrt{2}\ c_1 + c_2 = 0$$

$$c_1 + \sqrt{2}\ c_2 + c_3 = 0$$

$$c_2 + \sqrt{2}\ c_3 = 0$$

from which we obtain

$$c_1 = c_3 = -\ 1/\sqrt{2}\ c_2$$

Thus we have

$$\psi_3 = c_2\left[\frac{1}{\sqrt{2}}\chi_1 - \chi_2 + \frac{1}{\sqrt{2}}\chi_3\right]$$

Normalization gives

$$c_2^2\left[\frac{1}{2} + 1 + \frac{1}{2}\right] = 1$$

or

$$c_2 = 1/\sqrt{2}$$

Finally then we write

$$\psi_3 = \frac{1}{2}\chi_1 - \frac{1}{\sqrt{2}}\chi_2 + \frac{1}{2}\chi_3$$

For the allyl radical ($3\pi$ electrons)

$$E_\pi = 2(\alpha + \sqrt{2}\ \beta) + \alpha = 3\alpha + 2\sqrt{2}\ \beta$$

$$DE = (3\alpha + 2\sqrt{2}\ \beta) - (3\alpha + 2\beta) = 0.828\ \beta$$

We have subtracted $3\alpha + 2\beta$ from $E_\pi$ because the three localized $\pi$ electrons can be thought of as one ethylene molecule $(2\alpha + 2\beta)$ and a single carbon atom $(\alpha)$.

The charge on each carbon atom is given by (Eq. 9-153)

$$q_n = \sum_i n_i c_{in}^2$$

Thus

$$q_1 = 2\left[\frac{1}{2}\right]^2 + 1\left[\frac{1}{2}\right] = 1$$

$$q_2 = 2\left[\frac{1}{2}\right] = 1$$

$$q_3 = 2\left[\frac{1}{2}\right]^2 + 1\left[\frac{1}{2}\right] = 1$$

The $\pi$ bond orders are given by (Eq. 9-155)

$$P^{\pi}_{rs} = \sum_i n_i c_{ir} c_{is}$$

Thus

$$P^{\pi}_{12} = 2\left[\frac{\sqrt{2}}{4}\right] = 0.707$$

$$P^{\pi}_{23} = 2\left[\frac{\sqrt{2}}{4}\right] = 0.707$$

For the allyl carbonium ion (two $\pi$ electrons)

$$E_{\pi} = 2(\alpha + \sqrt{2}\beta) = 2\alpha + 2\sqrt{2}\ \beta$$

$$DE = E_{\pi} - 2(\alpha + \beta) = 0.828\ \beta$$

$$q_1 = 2\left[\frac{1}{2}\right]^2 = \frac{1}{2}$$

$$q_2 = 2\left[\frac{1}{2}\right] = 1$$

$$q_3 = 2\left[\frac{1}{2}\right]^2 = \frac{1}{2}$$

$$P^{\pi}_{12} = 2\left[\frac{\sqrt{2}}{4}\right] = 0.707$$

$$P^{\pi}_{23} = 2\left[\frac{\sqrt{2}}{4}\right] = 0.707$$

For the allyl carbanion (four $\pi$ electrons)

$$E_{\pi} = 2(\alpha + \sqrt{2}\beta) + 2\alpha = 4\alpha + 2\sqrt{2}\ \beta$$

$$DE = E_{\pi} - (4\alpha + 2\beta) = 0.828\ \beta$$

$$q_1 = 2\left[\frac{1}{2}\right]^2 + 2\left[\frac{1}{2}\right] = 1\frac{1}{2}$$

$$q_2 = 2\left[\frac{1}{2}\right] = 1$$

$$q_3 = 2\left[\frac{1}{2}\right]^2 + 2\left[\frac{1}{2}\right] = 1\frac{1}{2}$$

$$P^{\pi}_{12} = 2\left[\frac{\sqrt{2}}{4}\right] = 0.707$$

$$P^{\pi}_{23} = 2\left[\frac{\sqrt{2}}{4}\right] = 0.707$$

**9-41.** Because the simple Hückel molecular orbital theory that we discuss in Chapter 9 neglects overlap between adjacent carbon atoms, the normalization conditions of the molecular orbitals for benzene given in Example 9-13 are simply the sum of the squares of the coefficients of the $\chi_j$.
Thus, for $\psi_1$ we have

$$\left[\frac{1}{6} + \frac{1}{6} + \frac{1}{6} + \frac{1}{6} + \frac{1}{6} + \frac{1}{6}\right] = 1$$

For $\psi_2$ we have

$$\left[\frac{1}{4} + \frac{1}{4} + \frac{1}{4} + \frac{1}{4}\right] = 1$$

For $\psi_3$ we have

$$\left[\frac{1}{3} + \frac{1}{12} + \frac{1}{12} + \frac{1}{3} + \frac{1}{12} + \frac{1}{12}\right] = 1$$

etc. For the orthogonality conditions, we treat the molecular orbitals using the condition that the $\chi_j$ are orthonormal, that is,

$$\int \chi_i \chi_j = \delta_{ij}$$

Thus, for $\psi_1$ and $\psi_2$ we have

$$\int d\tau \psi_1 \psi_2 = \left[0 + \frac{1}{\sqrt{24}} + \frac{1}{\sqrt{24}} + 0 - \frac{1}{\sqrt{24}} - \frac{1}{\sqrt{24}}\right] = 0$$

For $\psi_1$ and $\psi_3$ we have

$$\int d\tau \psi_1 \psi_3 = \left[\frac{1}{\sqrt{18}} + \frac{1}{2\sqrt{18}} - \frac{1}{2\sqrt{18}} - \frac{1}{\sqrt{18}} - \frac{1}{2\sqrt{18}} + \frac{1}{2\sqrt{18}}\right] = 0$$

Note that the operation here is similar to taking the scalar product of two cartesian vectors in a six-dimensional space.

**9-42.** We number the carbon atoms in naphthalene as

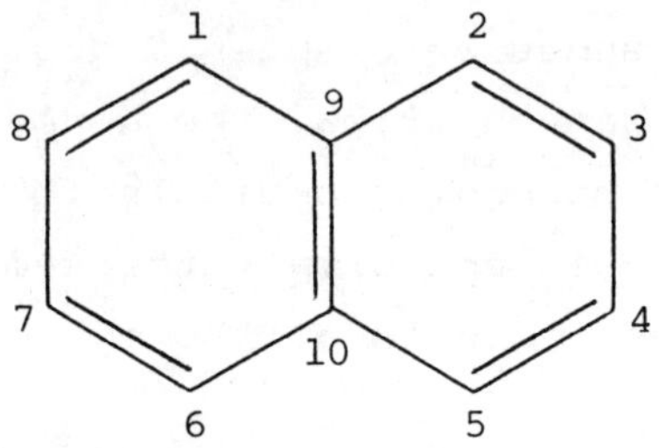


and so the Hückel determinant is

$$\begin{vmatrix} x & 0 & 0 & 0 & 0 & 0 & 0 & 1 & 1 & 0 \\ 0 & x & 1 & 0 & 0 & 0 & 0 & 0 & 1 & 0 \\ 0 & 1 & x & 1 & 0 & 0 & 0 & 0 & 0 & 0 \\ 0 & 0 & 1 & x & 1 & 0 & 0 & 0 & 0 & 0 \\ 0 & 0 & 0 & 1 & x & 0 & 0 & 0 & 0 & 1 \\ 0 & 0 & 0 & 0 & 0 & x & 1 & 0 & 0 & 1 \\ 0 & 0 & 0 & 0 & 0 & 1 & x & 1 & 0 & 0 \\ 1 & 0 & 0 & 0 & 0 & 0 & 1 & x & 0 & 0 \\ 1 & 1 & 0 & 0 & 0 & 0 & 0 & 0 & x & 1 \\ 0 & 0 & 0 & 0 & 1 & 1 & 0 & 0 & 1 & x \end{vmatrix}$$

**9-43.** There are ten $\pi$ electrons in naphthalene, and so the $\pi$-electron energy is given by

$$\begin{aligned} E_\pi &= 2(\alpha + 2.3028\beta) + 2(\alpha + 1.6180\beta) \\ &\qquad + 2(\alpha + 1.3029\beta) + 2(\alpha + \beta) \\ &\qquad + 2(\alpha + 0.6180\beta) \\ &= 10\alpha + 13.68\beta \end{aligned}$$

**9-44.** To calculate the delocalization energy of naphthalene, we subtract the energy of five ethylene molecules ($10\alpha + 10\beta$) from $E_\pi$ for naphthalene

$$\begin{aligned} \text{D.E.} &= 10\alpha + 13.68\beta - 10(\alpha + \beta) \\ &= 3.68\beta \end{aligned}$$

**9-45.** Use Eqs. (9-151).
For the butadiene cation (three $\pi$ electrons)

$$q_1 = q_4 = 2(0.3717)^2 + (0.6015)^2 = 0.6381$$

$$q_2 = q_3 = 2(0.6015)^2 + (0.3717)^2 = 0.8617$$

Note that $q_1 + q_2 + q_3 + q_4 = 3.00$
For the butadiene anion (five $\pi$ electrons)

$$\begin{aligned} q_1 = q_4 &= 2(0.3717)^2 + 2(0.6015)^2 + (0.6015)^2 \\ &= 1.362 \end{aligned}$$

$$\begin{aligned} q_2 = q_3 &= 2(0.6015)^2 + 2(0.3717)^2 + (0.3717)^2 \\ &= 1.138 \end{aligned}$$

Note that $q_1 + q_2 + q_3 + q_4 = 5.00$

**9-46.** The $\pi$-electron energy is given by

$$E_\pi = \sum_i n_i E_i$$

where $n_i$ is the number of electrons in the ith molecular orbital and $E_i$ is its energy. The energy of the ith molecular orbital is given by

$$E_i = \int \psi_i \hat{H}_{eff} \psi_i d\tau$$

where (p. 419)

$$\psi_i = \sum_r c_{ir} \chi_r$$

Substituting $\psi_i$ into the integral for $E_i$ yields

$$\begin{aligned} E_i &= \int \psi_i \hat{H}_{eff} \psi_i d\tau = \int d\tau \left[\sum_r c_{ir}\chi_r\right] \hat{H}_{eff} \left[\sum_s c_{is}\chi_s\right] \\ &= \sum_r \sum_s c_{ir} c_{is} \int d\tau \chi_r \hat{H}_{eff} \chi_s = \sum_r \sum_s c_{ir} c_{is} H_{rs} \end{aligned}$$

We now let $H_{rr} = \alpha_r$ and $H_{rs} = \beta_{rs}$ $(r \neq s)$ and write the double summation in the form

$$E_i = \sum_r c_{ir}^2 \alpha_r + \sum_r \sum_s c_{ir} c_{is} \beta_{rs} \quad r \neq s$$

Now subsitute this expression for $E_i$ into the expression for $E_\pi$ to get

$$E_\pi = \sum_i n_i E_i = \sum_i \sum_r n_i c_{ir}^2 \alpha_r + \sum_i \sum_r \sum_s n_i c_{ir} c_{is} \beta_{rs} \quad r \neq s$$

Use Eq. 9-153 in the first summation and Eq. 9-155 in the second summation to get

$$E_\pi = \sum_r q_r \alpha_r + \sum_r \sum_s P_{rs}^\pi \beta_{rs} \quad r \neq s$$

**9-47.** For the triangular geometry, the Hückel determinantal equation is

$$\begin{vmatrix} x & 1 & 1 \\ 1 & x & 1 \\ 1 & 1 & x \end{vmatrix} = x^3 - 3x + 2 = 0$$

The three roots of this equation are

$$x = 1,\ 1,\ -2$$

The three energies are

$$E = \begin{cases} \alpha - \beta & \text{(twice)} \\ \alpha + 2\beta \end{cases}$$

The energy of $H_3^+$ (two electrons) is

$$E_{H_3^+} = 2(\alpha + 2\beta) = 2\alpha + 4\beta$$

The energy of $H_3$ (three electrons) is

$$E_{H_3} = 2(\alpha + 2\beta) + \alpha - \beta = 3\alpha + 3\beta$$

and the energy of $H_3^-$ (four electrons) is

$$E_{H_3^-} = 2(\alpha + 2\beta) + 2(\alpha - \beta) = 4\alpha + 2\beta$$

For the linear geometry, the Hückel determinantal equation is

$$\begin{vmatrix} x & 1 & 0 \\ 1 & x & 1 \\ 0 & 1 & x \end{vmatrix} = x^3 - 2x = 0$$

The three roots of this equation are

$$x = 0, \pm \sqrt{2}$$

The energies are

$$E = \begin{cases} \alpha - \sqrt{2}\beta \\ \alpha \\ \alpha + \sqrt{2}\beta \end{cases}$$

and the energies of $H_3^+$ (two electrons), $H_3$ (three electrons) and $H_3^-$ (four electrons) are

$$E_{H_3^+} = 2(\alpha + \sqrt{2}\beta) = 2\alpha + 2\sqrt{2}\beta$$

$$E_{H_3} = 2(\alpha + \sqrt{2}\beta) + \alpha = 3\alpha + 2\sqrt{2}\beta$$

$$E_{H_3^-} = 2(\alpha + \sqrt{2}\beta) + 2\alpha = 4\alpha + 2\sqrt{2}\beta$$

The triangular geometry is more stable for $H_3^+$ ($2\alpha + 4\beta$ vs. $2\alpha + 2\sqrt{2}\beta$), the linear geometry is more stable for $H_3^-$ ($4\alpha + 2\beta$ vs. $4\alpha + 2\sqrt{2}\beta$), and the two geometries have almost the same energy for $H_3$ ($3\alpha + 3\beta$ vs. $3\alpha + 2\sqrt{2}\beta$).

**9-48.** Use the following formula of pyridine to set up the Hückel theory secular determinant

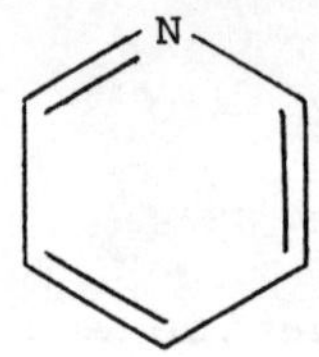

The Hückel theory secular determinant is

$$\begin{vmatrix} \alpha_N - E & \beta_{CN} & 0 & 0 & 0 & \beta_{CN} \\ \beta_{CN} & \alpha_C - E & \beta_{CC} & 0 & 0 & 0 \\ 0 & \beta_{CC} & \alpha_C - E & \beta_{CC} & 0 & 0 \\ 0 & 0 & \beta_{CC} & \alpha_C - E & \beta_{CC} & 0 \\ 0 & 0 & 0 & \beta_{CC} & \alpha_C - E & \beta_{CC} \\ \beta_{CN} & 0 & 0 & 0 & \beta_{CC} & \alpha_C - E \end{vmatrix}$$

**9-49.** The force constant is the curvature of the internuclear potential at its minimum. Because the masses of the nuclei are considered to be infinite under the Born-Oppenheimer approximation, the internuclear potential, and hence the force constant, is independent of isotopic substitution. Because

$$\nu = \frac{1}{2\pi}\left[\frac{k}{\mu}\right]^{1/2}$$

we can write

$$\frac{\nu_{D_2}}{\nu_{H_2}} = \left[\frac{\mu_{H_2}}{\mu_{D_2}}\right]^{1/2} = \frac{(1/2)^{1/2}}{(1)^{1/2}} = 0.707$$

Thus

$$\nu_{D_2} = 0.707\nu_{H_2} = (0.707)(1.32 \times 10^{14}\ \text{s}^{-1})$$
$$= 9.33 \times 10^{13}\ \text{s}^{-1}$$

Furthermore, the value of $D_e$ is independent of isotopic substitution, and so we can write

$$D_e = D_{o,H_2} + \frac{1}{2}h\nu_{H_2} = D_{o,D_2} + \frac{1}{2}h\nu_{D_2}$$

Therefore

$$\begin{aligned} D_{o,D_2} &= D_{o,H_2} + \frac{1}{2}h(\nu_{H_2} - \nu_{D_2}) \\ &= (4.46\ \text{eV})(1.602 \times 10^{-19}\ \text{J}\cdot\text{eV}^{-1}) \\ &\qquad + [(1/2)(6.626 \times 10^{-34}\ \text{J}\cdot\text{s}) \\ &\qquad\qquad \times (3.87 \times 10^{13}\ \text{s}^{-1})] \\ &= 7.27 \times 10^{-19}\ \text{J} \\ &= 4.54\ \text{eV} \end{aligned}$$

# CHAPTER TEN

# MOLECULAR SPECTROSCOPY

**10-1.** From Eq. 10-14, the spacing is given by

$$\text{spacing} = \frac{h}{4\pi^2 I} = 6.350 \times 10^{11}\ \text{s}^{-1}$$

$$I = \frac{(6.626 \times 10^{-34}\ \text{J}\cdot\text{s})}{4\pi^2(6.350 \times 10^{11}\ \text{s}^{-1})} = 2.643 \times 10^{-47}\ \text{kg}\cdot\text{m}^2$$

$$R_o = \left[\frac{I}{\mu}\right]^{1/2}$$

$$= \left[\frac{2.643 \times 10^{-47}\ \text{kg}\cdot\text{m}^2}{(35/36\ \text{amu})(1.6606 \times 10^{-27}\ \text{kg}\cdot\text{amu}^{-1})}\right]^{1/2}$$

$$= 1.28 \times 10^{-10}\ \text{m} = 128\ \text{pm}$$

**10-2.** From Eq. 10-14, the spacing is given by

$$\text{spacing} = \frac{h}{4\pi^2 I} = 3\ 634 \times 10^{6}\ \text{s}^{-1}$$

$$I = \frac{(6.626 \times 10^{-34}\ \text{J}\cdot\text{s})}{4\pi^2(3.634 \times 10^{9}\ \text{s}^{-1})} = 4.619 \times 10^{-45}\ \text{kg}\cdot\text{m}^2$$

$$R_o = \left[\frac{I}{\mu}\right]^{1/2}$$

$$= \left[\frac{4.619 \times 10^{-45}\ \text{kg}\cdot\text{m}^2}{\left[\frac{39 \times 127}{39 + 127}\text{amu}\right](1.6606 \times 10^{-27}\ \text{kg}\cdot\text{amu}^{-1})}\right]^{1/2}$$

$$= 3.05 \times 10^{-10}\ \text{m} = 305\ \text{pm}$$

**10-3.** From Eq. 10-16,

$$\bar{B} = \frac{h}{8\pi^2 c I}$$

The moment of inertia of $H^{127}I$ is

$$I = \mu r^2 = \left[\frac{127}{128}\text{amu}\right](1.6606 \times 10^{-27}\ \text{kg}\cdot\text{amu}^{-1}) \times (1.604 \times 10^{-10}\text{m})^2$$

$$= 4.239 \times 10^{-47}\ \text{kg}\cdot\text{m}^2$$

Therefore,

$$\bar{B} = \frac{h}{8\pi^2 c I}$$

$$= \frac{(6.626 \times 10^{-34}\ \text{J·s})}{8\pi^2(2.998 \times 10^8\ \text{m·s}^{-1})(4.239 \times 10^{-47}\ \text{kg·m}^2)}$$

$$= 660\ \text{m}^{-1} = 6.60\ \text{cm}^{-1}$$

$$B = c\bar{B} = (2.998 \times 10^8\ \text{m·s}^{-1})(660\ \text{m}^{-1})$$

$$= 1.98 \times 10^{11}\ \text{s}^{-1}$$

$$= 1.98 \times 10^5\ \text{MHz}$$

**10-4.** The energy of a classical rotator is

$$K = (1/2)I\omega^2$$

The energy of a rotator in the Jth rotational state is

$$E = hBJ(J + 1)$$

By equating K and E, we have

$$\omega = \left[\frac{2hBJ(J + 1)}{I}\right]^{1/2} = 4\pi B\sqrt{J(J + 1)}$$

$$= 4\pi(6500 \times 10^6\ \text{s}^{-1})(110)^{1/2}$$

$$= 8.57 \times 10^{11}\ \text{radian·s}^{-1}$$

$$= 1.36 \times 10^{11}\ \text{revolutions·s}^{-1}$$

**10-5.** The microwave spectrum of $H^{12}C^{14}N$ consists of a series of equally spaced lines with separation

$$2\bar{B} = \frac{h}{4\pi^2 cI}$$

$$= \frac{6.626 \times 10^{-34}\ \text{J·s}}{4\pi^2(2.998 \times 10^8\ \text{m·s}^{-1})(1.89 \times 10^{-46}\ \text{kg·m}^2)}$$

$$= 296\ \text{m}^{-1} = 2.96\ \text{cm}^{-1} = 8.88 \times 10^{10}\ \text{s}^{-1}$$

**10-6.** The easiest way to do this problem is to work backwards. Start with

$$MI = m_H m_C(r_H - r_C)^2 + m_H m_N(r_H - r_N)^2 + m_C m_N(r_C - r_N)^2$$

where $M = m_H + m_C + m_N$.

$$MI = m_H m_C r_H^2 - 2m_H m_C r_H r_C + m_H m_C r_C^2 + m_H m_N m_H^2 - 2m_H m_N r_H r_N + m_H m_N r_N^2 + m_C m_N r_C^2 - 2m_C m_N r_C r_N + m_C m_N r_N^2$$

$$= m_H(m_C + m_N)r_H^2 + m_C(m_H + m_N)r_C^2$$
$$+ m_N(m_H + m_C)r_N^2 - 2m_Hm_Cr_Hr_C$$
$$- 2m_Hm_Nr_Hr_N - 2rm_Cm_Nr_Cr_N$$

Now add and subtract $m_Hr_H^2 + m_Cr_C^2 + m_Nr_N^2$

$$MI = Mm_Hr_H^2 + Mm_Cr_C^2 + Mm_Nr_N^2$$
$$- \left[m_Hr_H^2 + m_Cr_C^2 + m_Nr_N^2 + 2m_Hm_Cr_Hr_C + 2m_Hm_Nr_Hr_N + 2m_Cm_Nr_Cr_N\right]$$

$$= M(m_Hr_H^2 + m_Cr_C^2 + m_Nr_N^2) - \left[m_Hr_H + m_Cr_C + m_Nr_N\right]^2$$

The term in brackets equals zero by definition of the center of mass, and so we finally have

$$I = m_Hr_H^2 + m_Cr_C^2 + m_Nr_N^2$$

Thus, the two expressions for I are equivalent to each other.

Numerically,

$$I = \frac{\begin{array}{l}(1.00\text{ amu})(12.00\text{ amu})(106.8\text{ pm})^2 \\ \quad + (1.00\text{ amu})(14.00\text{amu}) \\ \quad\quad \times (106.8\text{ pm} + 115.6\text{ pm})^2 \\ \quad + (12.00\text{ amu})(14.00\text{ amu})(115.6\text{ pm})^2\end{array}}{(1.00) + 12.00 + 14.00)\text{amu}}$$

$$= \frac{3.074 \times 10^6\text{ amu}^2\cdot\text{pm}^2}{27.00\text{ pm}} = 1.138 \times 10^5\text{ amu}\cdot\text{pm}^2$$

$$= 1.138 \times 10^5\text{ amu}\cdot\text{pm}^2\left[\frac{1.6607\times10^{-27}\text{kg}}{\text{amu}}\right]\left[\frac{10^{-12}\text{m}}{\text{pm}}\right]^2$$

$$= 1.890 \times 10^{-46}\text{ kg}\cdot\text{m}^2$$

This is the same value given in Problem 10-5.

**10-7.** Use Eq. 10-21.

$$\nu_{obs} = \frac{1}{2\pi}\left[\frac{k}{\mu}\right]^{1/2}$$

or

$$\bar{\nu}_{obs} = \frac{1}{2\pi c}\left[\frac{k}{\mu}\right]^{1/2}$$

$$k = (2\pi c\bar{\nu}_{obs})^2\mu$$

$$= [2\pi(2.998 \times 10^8\ \text{m}\cdot\text{s}^{-1})(378.0 \times 10^2\ \text{m}^{-1})]^2 \times \left[\frac{39\times 35}{74}\text{amu}\right](1.6606 \times 10^{-27}\ \text{kg}\cdot\text{amu}^{-1})$$

$$= 155\ \text{N}\cdot\text{m}^{-1}$$

The period of vibration is

$$T = \frac{1}{\nu_{obs}} = \frac{1}{c\bar{\nu}_{obs}} = 8.82 \times 10^{-14}\ \text{s}$$

**10-8.** Use Eq. 10-21

$$\nu_{obs} = \frac{1}{2\pi}\left[\frac{k}{\mu}\right]^{1/2}$$

$$= \frac{1}{2\pi}\left[\frac{240\ \text{N}\cdot\text{m}^{-1}}{\left[\frac{79\times 79}{158}\text{amu}\right](1.6606\times 10^{-27}\ \text{kg}\cdot\text{amu}^{-1})}\right]^{1/2}$$

$$= 9.62 \times 10^{12}\ \text{s}^{-1}$$

$$= 321\ \text{cm}^{-1}$$

$$E_0 = (1/2)h\nu_{obs} = 3.19 \times 10^{-21}\ \text{J}$$

**10-9.** The proof is given on p. 183

$$x_{rms} = (\langle x^2\rangle)^{1/2} = \left[\frac{\hbar}{2(\mu k)^{1/2}}\right]^{1/2}$$

$$= \left[\frac{1.055 \times 10^{-34}\ \text{J}\cdot\text{s}}{2\left[\begin{array}{c}((14 \times 14)/28\ \text{amu}) \\ \times (1.6606 \times 10^{-27}\ \text{kg}\cdot\text{amu}^{-1}) \\ \times (2260\ \text{N}\cdot\text{m}^{-1})\end{array}\right]^{1/2}}\right]^{1/2}$$

$$= 3.21 \times 10^{-12}\ \text{m} = 3.21\ \text{pm}$$

**10-10.** Use Eq. (10-25).

| 300K state | f | 1000K state | f |
|---|---|---|---|
| 0 | ~1 | 0 | 0.965 |
| 1 | $1.4 \times 10^{-5}$ | 1 | $3.4 \times 10^{-2}$ |
| 2 | $1.9 \times 10^{-10}$ | 2 | $1.2 \times 10^{-3}$ |

**10-11.** For $H_2$,

$$f_0 = 1 - \exp(-h\nu/k_BT)$$

$$= 1 - \exp\left[-\frac{(6.626\times10^{-34}\text{ J·s})(4159.2\text{ cm}^{-1})\times(2.998\times10^{10}\text{ cm·s}^{-1})}{(1.38066\times10^{-23}\text{ J·K}^{-1})(300\text{ K})}\right]$$

$$\approx 1$$

For NaCl,

$$f_0 = 1 - \exp\left[-\frac{(6.626\times10^{-34}\text{ J·s})(378.0\text{ cm}^{-1})\times(2.998\times10^{10}\text{ cm·s}^{-1})}{(1.38066\times10^{-23}\text{ J·K}^{-1})(300\text{ K})}\right]$$

$$= 0.84$$

**10-12.** From Eqs. 10-29 and 10-30,

$$\nu_R = \nu_0 + 2c\bar{B}(J+1) \qquad J = 0, 1, 2, \ldots$$

$$\nu_P = \nu_0 - 2c\bar{B}J \qquad J = 1, 2, \ldots$$

or

$$\bar{\nu}_R = 2143.0\text{ cm}^{-1} + 2\left[\frac{56{,}000\times10^{6}\text{ s}^{-1}}{2.998\times10^{10}\text{ cm·s}^{-1}}\right](J+1)$$

$$= 2143.0\text{ cm}^{-1} + (3.74\text{ cm}^{-1})(J+1)$$

$$\bar{\nu}_P = 2143.0\text{ cm}^{-1} - (3.74\text{ cm}^{-1})J$$

**10-13.**

$$I = \mu R_0^2 = \left[\frac{6\times19}{25}\text{amu}\right](1.6606\times10^{-27}\text{ kg·amu}^{-1})\times(156\times10^{-12}\text{ m})^2$$

$$= 1.843\times10^{-46}\text{ kg·m}^2$$

$$\bar{B} = \frac{h}{8\pi^2cI}$$

$$= \frac{(6.626\times10^{-34}\text{ J·s})}{8\pi^2(2.998\times10^{8}\text{ m·s}^{-1})(1.843\times10^{-46}\text{ kg·m}^2)}$$

$$= 151.9\text{ m}^{-1} = 1.52\text{ cm}^{-1}$$

$$\nu_0 = \frac{1}{2\pi}\left[\frac{k}{\mu}\right]^{1/2}$$

$$= \frac{1}{2\pi}\left[\frac{250\text{ N·m}^{-1}}{((6\times19)/25\text{ amu})\times(1.6606\times10^{-27}\text{ kg·amu}^{-1})}\right]^{1/2}$$

$$= 2.891\times10^{13}\text{ s}^{-1}$$

and

$$\bar{\nu}_0 = \frac{\nu_0}{c} = 964.6\ \text{cm}^{-1}$$

Therefore,

$$\begin{aligned}\bar{\nu}_R &= \bar{\nu}_0 + 2\bar{B}(J + 1)\\ &= 964.6\ \text{cm}^{-1} + (3.04\ \text{cm}^{-1})(J + 1)\\ \bar{\nu}_P &= \bar{\nu}_0 - 2\bar{B}J\\ &= 964.6\ \text{cm}^{-1} - (3.04\ \text{cm}^{-1})J\end{aligned}$$

**10–14.** From Eq. 10–33,

$$\frac{N_J}{N_0} = (2J + 1)\exp\left[-\frac{\bar{B}J(J + 1)}{k_BT}\right]$$

$$\bar{B} = \frac{B}{c} = \frac{(56{,}000 \times 10^6\ \text{s}^{-1})}{(2.998 \times 10^8\ \text{m}\cdot\text{s}^{-1})}$$

$$= 187\ \text{m}^{-1} = 1.87\ \text{cm}^{-1}$$

$$\frac{N_J}{N_0} = (2J + 1)\exp\left[-\frac{(1.87\ \text{cm}^{-1})(J(J + 1)}{(0.69509\ \text{cm}^{-1}\cdot\text{K}^{-1})(298\ \text{K})}\right]$$

$$= (2J + 1)\exp(-0.00903\ J(J + 1))$$

To find $J_{max}$,

$$\frac{d}{dJ}\left[\frac{N_J}{N_0}\right] = 0$$

$$\rightarrow 2\exp(-\bar{B}J(J + 1)/k_BT) - \frac{\bar{B}}{k_BT}(2J + 1)^2\exp(-\bar{B}J(J + 1)/k_BT)$$

$$= 0$$

Thus

$$(2J_{max} + 1)^2 = \frac{2k_BT}{\bar{B}}$$

or

$$J_{max} = \frac{1}{2}\left[\left[\frac{2k_BT}{\bar{B}}\right]^{1/2} - 1\right]$$

For $H^{35}Cl$,

$$J_{max} = \frac{1}{2}\left[\left[\frac{2 \times 0.69509\ \text{cm}^{-1}\cdot\text{K}^{-1} \times 300\text{K}}{10.60\ \text{cm}^{-1}}\right]^{1/2} - 1\right]$$

$$= 2.6\ (J_{max}\ \text{is between 2 and 3})$$

For $^{127}I^{35}Cl$,

$$J_{max} = \frac{1}{2}\left[\left[\frac{2 \times 0.69509\ \text{cm}^{-1}\cdot\text{K}^{-1} \times 300\text{K}}{0.114\ \text{cm}^{-1}}\right]^{1/2} - 1\right]$$

$$= 29.7 \quad (J_{max} \approx 30)$$

**10-15.**

$$N_J = N_o(2J + 1)\exp(-\bar{B}J(J + 1)/k_BT)$$

$$\sum_{J=o}^{\infty} N_J = N = N_o \sum_{J=o}^{\infty} (2J + 1)\exp(-\bar{B}J(J + 1)/k_BT)$$

$$\approx N_o\left[\frac{k_BT}{\bar{B}}\right]$$

$$f_J = \frac{N_J}{N} = \frac{(2J + 1)}{(k_BT/\bar{B})}\exp(-\bar{B}J(J + 1)/k_BT)$$

**10-16.** From Eq. 10-38,

$$\text{ratio} = \frac{\bar{D}J^2(J + 1)^2}{\bar{B}J(J + 1) - \bar{D}J^2(J + 1)^2}$$

$$= \frac{\bar{D}J(J + 1)/\bar{B}}{1 - \bar{D}J(J + 1)/\bar{B}}$$

$$= \frac{(0.0005\ \text{cm}^{-1})(10)(11)/10.6\ \text{cm}^{-1}}{1 - (0.0005\ \text{cm}^{-1})(10)(11)/10.6\ \text{cm}^{-1}}$$

$$= 0.005$$

**10-17.**

$$\bar{\nu} = 2\bar{B}(J + 1) - 4\bar{D}(J + 1)^3$$

$$\frac{\bar{\nu}}{J + 1} = 2\bar{B} - 4\bar{D}(J + 1)^2$$

The data from Table 10-2 are plotted in the following figure. (Compare the accuracy in this figure to that in Problem 10-18.)

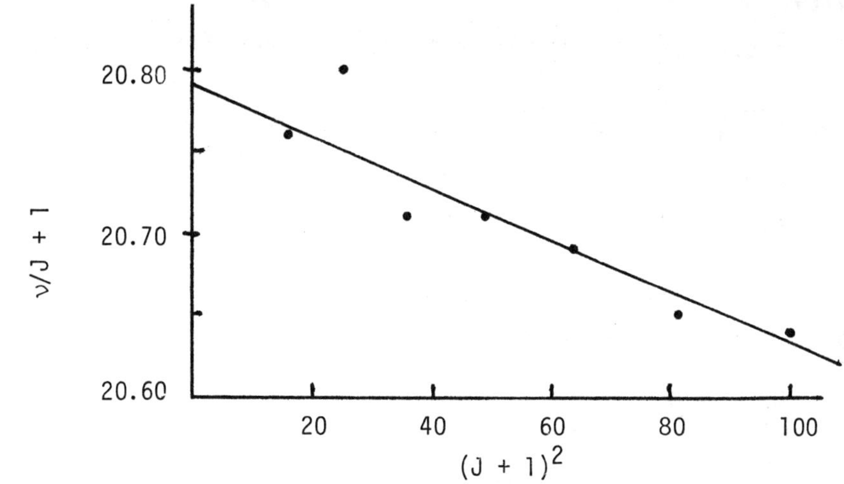


The intercept gives $2\bar{B} = 20.79\ cm^{-1}$ and the slope gives $4\bar{D} = 0.0016\ cm^{-1}$.

**10-18.** The data are plotted in the following figure

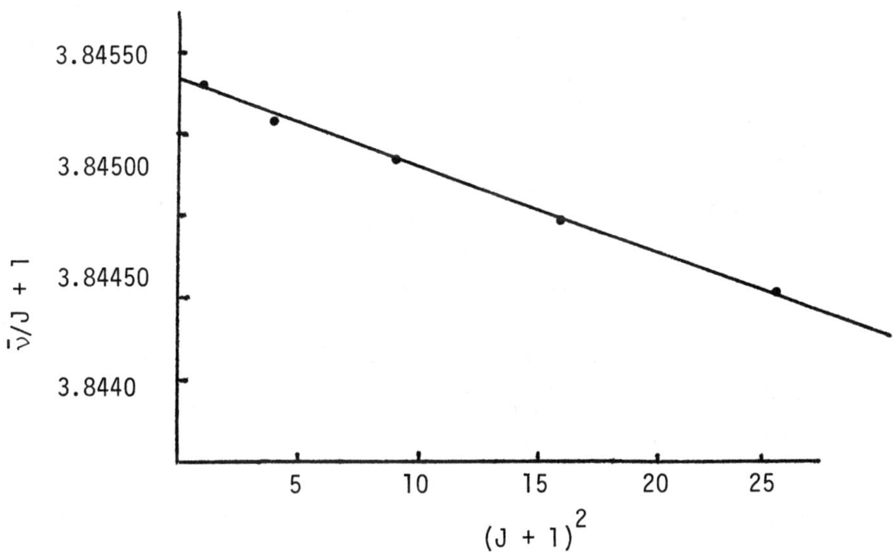


The intercept gives $2\bar{B} = 3.8454\ cm^{-1}$ and the slope gives $4\bar{D} = 0.00065\ cm^{-1}$

10-19.

$$\bar{\nu}_{calc}/cm^{-1} = 20.790(J + 1) - 0.0016(J + 1)^3$$

| | $\bar{\nu}_{calc}/cm^{-1}$ |
|---|---|
| $0 \to 1$ | 20.788 |
| $1 \to 2$ | 41.567 |
| $2 \to 3$ | 62.327 |
| $3 \to 4$ | 83.057 |

10-20. Start with (p. 450)

$$\frac{1}{R^2} = \frac{1}{R_0^2}\left[1 - \frac{2L^2}{\mu k R_0^4} + \frac{L^4}{\mu^2 k^2 R_0^8}\right]$$

or

$$R^2 = R_0^2\left[1 - \frac{2L^2}{\mu k R_0^4} + \frac{L^4}{\mu^2 k^2 R_0^8}\right]^{-1}$$

Replace $L^2$ by $\hbar J(J + 1)$, take the square root and use the expansion

$$(1 + x)^{-1/2} = 1 - (1/2)x + \cdots$$

to get

$$R = R_0\left[1 + \frac{\hbar^2 J(J + 1)}{\mu k R_0^4} + \cdots\right]$$

$$= R_0 + \frac{\hbar^2 J(J + 1)}{\mu k R_0^3} + \cdots$$

10-21.

$$\text{fraction} = \frac{\bar{x}_e\bar{\nu}_e(n + 1/2)^2}{\bar{\nu}_e(n + 1/2) - \bar{x}_e\bar{\nu}_e(n + 1/2)^2}$$

$$= \frac{52.05(n + 1/2)}{2987.7 - 52.05(n + 1/2)}$$

| n | fraction |
|---|---|
| 0 | 0.0088 |
| 5 | 0.1059 |
| 10 | 0.2237 |

10-22. Expand $U(q)$ in a Maclaurin series

$$U(q) = D_e(1 - e^{-\beta q})^2$$

$$= D_e[1 - (1 - \beta q + (1/2)\beta^2 q^2 + \cdots)]^2$$

$$= D_e(\beta^2 q^2 + \cdots)$$

Comparing this result to

$$U(q) = (1/2)kq^2$$

gives

$$k = 2D_e\beta^2$$
$$= 2(7.31\times10^{-19}\ \text{J}\cdot\text{molecule}^{-1})(1.83\times10^{10}\ \text{m}^{-1})^2$$
$$= 490\ \text{N}\cdot\text{m}^{-1}$$

**10-23.**

$$\bar{\nu}_e = \frac{1}{2\pi c}\left[\frac{k}{\mu}\right]^{1/2}$$

$$k = (2\pi c\bar{\nu}_e)^2\mu$$
$$= (2\pi \times 2.998 \times 10^{10}\ \text{cm}\cdot\text{s}^{-1} \times 1580.0\ \text{cm}^{-1})^2 \times (8\ \text{amu})(1.6606 \times 10^{-27}\ \text{kg}\cdot\text{amu}^{-1})$$
$$= 1180\ \text{N}\cdot\text{m}^{-1}$$

$$\beta = \left[\frac{k}{2D_e}\right]^{1/2} = \left[\frac{1180\ \text{N}\cdot\text{m}^{-1}}{2 \times 8.16 \times 10^{-19}\ \text{J}}\right]^{1/2}$$
$$= 2.69 \times 10^{10}\ \text{m}^{-1}$$

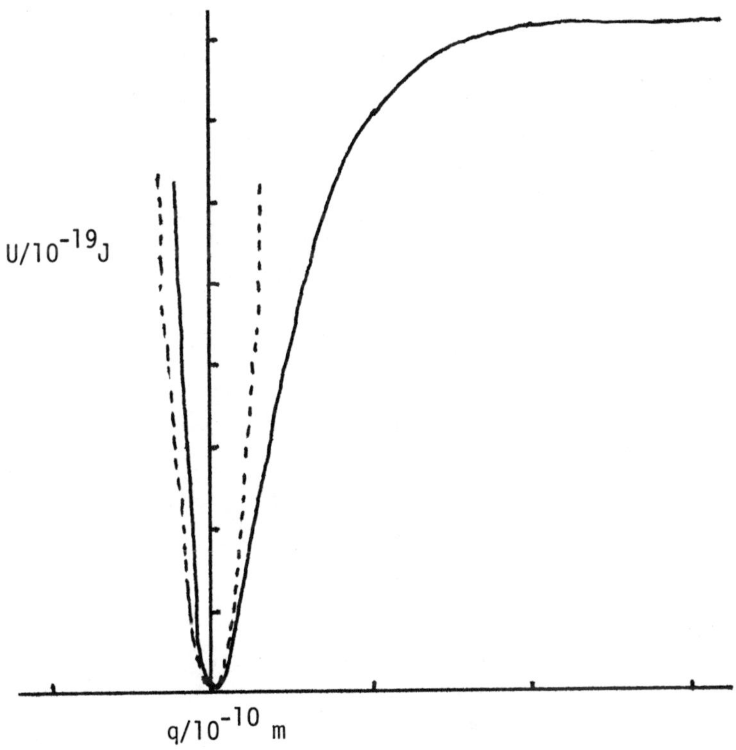

**10-24.** From Example 10-8,

fundamental: $\bar{\nu}_{obs} = \bar{\nu}_e - 2\bar{x}_e\bar{\nu}_e = 2143.0\ \text{cm}^{-1}$

first overtone: $\bar{\nu}_{obs} = 2\bar{\nu}_e - 6\bar{x}_e\bar{\nu}_e = 4260.0\ \text{cm}^{-1}$

Multiply the fundamental frequency by 3 and subtract to get

$$\bar{\nu}_e = 3(2143.0\ \text{cm}^{-1}) - 4260.0\ \text{cm}^{-1} = 2169\ \text{cm}^{-1}$$

Multiply the fundamental frequency by 2 and subtract to get

$$2\bar{x}_e\bar{\nu}_e = 26.0\ \text{cm}^{-1} \longrightarrow \bar{x}_e\bar{\nu}_e = 13.0\ \text{cm}^{-1}$$

**10-25.** From Example 10-8,

fundamental: $\bar{\nu}_{obs} = \bar{\nu}_e - 2\bar{x}_e\bar{\nu}_e$

$$= 2649.7\ \text{cm}^{-1} - 2(45.21\ \text{cm}^{-1})$$

$$= 2559.3\ \text{cm}^{-1}$$

first overtone: $\bar{\nu}_{obs} = 2(2649.7\ \text{cm}^{-1})$

$$- 6(45.21\ \text{cm}^{-1})$$

$$= 5028.1\ \text{cm}^{-1}$$

second overtone: $\bar{\nu}_{obs} = 3(2649.7\ \text{cm}^{-1})$

$$- 12(45.21\ \text{cm}^{-1})$$

$$= 7406.6\ \text{cm}^{-1}$$

third overtone: $\bar{\nu}_{obs} = 4(2649.7\ \text{cm}^{-1})$

$$- 20(45.21\ \text{cm}^{-1})$$

$$= 9694.6\ \text{cm}^{-1}$$

**10-26.**

$$\bar{\nu}_{obs} = \bar{\nu}_e n - \bar{x}_e\bar{\nu}_e n(n + 1)$$

$$\bar{\nu}_{obs}/n = \bar{\nu}_e - \bar{x}_e\bar{\nu}_e(n + 1)$$

The data are plotted in the following figure

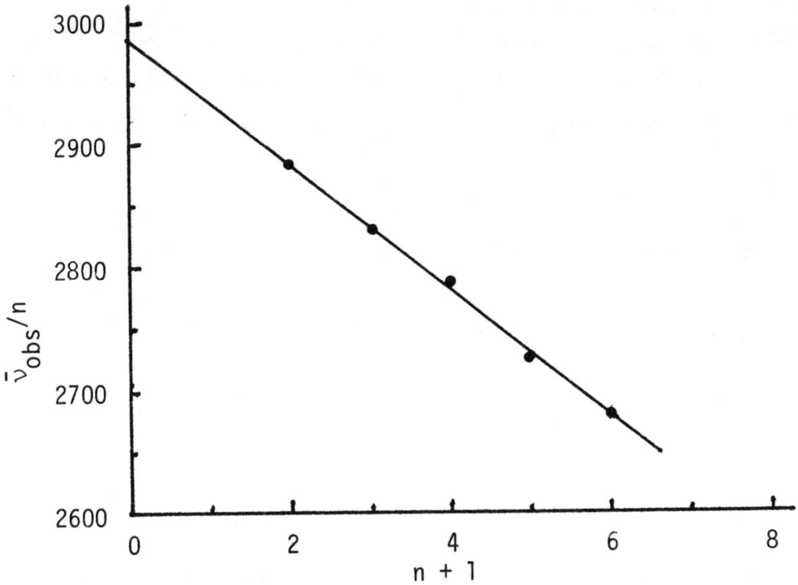


The intercept gives $\bar{\nu}_e$ = 2990 cm$^{-1}$ and the slope gives $\bar{x}_e\bar{\nu}_e$ = 51.7 cm$^{-1}$

**10-27.** The data are plotted in the following figure

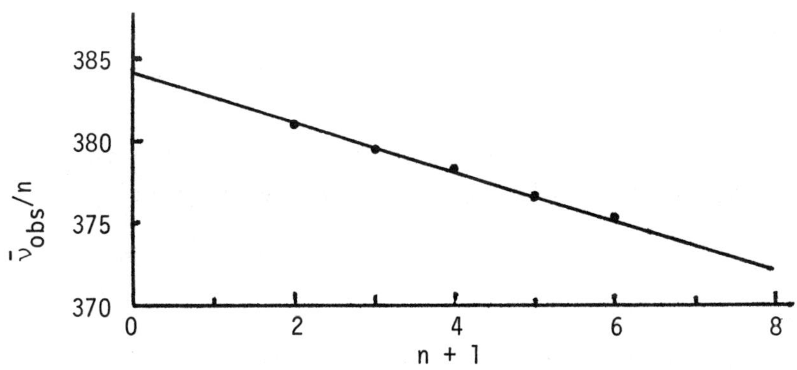


The intercept gives $\bar{\nu}_e$ = 3842.2 cm$^{-1}$ and the slope gives $\bar{x}_e\bar{\nu}_e$ = 1.48 cm$^{-1}$.

**10-28.** As n increases in Figure 10-6, the width of the "box" increases, and so the energy levels bunch up because $E \sim 1/a^2$ for a particle in a box.

**10-29.** Use the equation

$$\bar{E}_{nJ} = \bar{\nu}_e(n + 1/2) - \bar{x}_e\bar{\nu}_e(n + 1/2)^2 + \bar{B}_n J(J + 1)$$

where $\bar{B}_n = \bar{B}_e - \bar{\alpha}_e(n + 1/2)$. For $H^{35}Cl$

$$\bar{E}_{nJ}/\text{cm}^{-1} = 2989.7(n + 1/2) - 0.0005(n + 1/2)^2 + 10.591\ J(J + 1) - 0.3019(n + 1/2)J(J + 1)$$

The energies are

| n | J | $\bar{E}_{nJ}/\text{cm}^{-1}$ |
|---|---|---|
| 0 | 0 | 1494.8 |
| | 1 | 1515.7 |
| | 2 | 1557.4 |
| | 3 | 1620.1 |
| | 4 | 1703.6 |
| 1 | 0 | 4484.5 |
| | 1 | 4504.8 |
| | 2 | 4545.3 |
| | 3 | 4606.2 |
| | 4 | 4687.3 |

For absorption in the infrared region, the allowed transitions are $\Delta n = +1$ and $\Delta J = \pm 1$. The frequencies of these transitions are

$$\bar{\nu}_R(\Delta J = +1) = \bar{\nu}_e - 2\bar{x}_e\bar{\nu}_e + 2\bar{B}_1 + (3\bar{B}_1 - \bar{B}_0)J + (\bar{B}_1 - \bar{B}_0)J^2 \qquad J = 0,1,2,\ldots$$

$$\bar{\nu}_P(\Delta J = -1) = \bar{\nu}_e - 2\bar{x}_e\bar{\nu}_e - (\bar{B}_1 + \bar{B}_0)J + (\bar{B}_1 - \bar{B}_0)J^2 \qquad J = 1,2,3,\ldots$$

or

$$\bar{\nu}_R/\text{cm}^{-1} = 2885.6 + 20.28 + 19.97\ J - 0.30\ J^2$$

$$\bar{\nu}_P/\text{cm}^{-1} = 2885.6 - 20.58\ J - 0.30\ J^2$$

**10-30.** Following Example 10-9, we have

$$R(0) = \bar{\nu}_0 + 2\bar{B}_1 = 2242.6 \text{ cm}^{-1}$$

$$R(1) = \bar{\nu}_0 + 6\bar{B}_1 - 2\bar{B}_0 = 2254.8 \text{ cm}^{-1}$$

$$P(1) = \bar{\nu}_0 - 2\bar{B}_0 = 2217.1 \text{ cm}^{-1}$$

$$P(2) = \bar{\nu}_0 + 2\bar{B}_1 - 6\bar{B}_0 = 2203.8 \text{ cm}^{-1}$$

$$R(1) - P(1) = 6\bar{B}_1 = 37.7 \text{ cm}^{-1} \rightarrow \bar{B}_1 = 6.28 \text{ cm}^{-1}$$

$$R(0) - P(2) = 6\bar{B}_0 = 38.8 \text{ cm}^{-1} \rightarrow \bar{B}_0 = 6.47 \text{ cm}^{-1}$$

$$\bar{B}_0 - \bar{B}_1 = \bar{B}_e - (1/2)\bar{\alpha}_e - (\bar{B}_e - (3/2)\bar{\alpha}_e)$$

$$= \bar{\alpha}_e = 0.19 \text{ cm}^{-1}$$

$$\bar{B}_e = \bar{B}_0 + (1/2)\bar{\alpha}_e = 6.56 \text{ cm}^{-1}$$

**10-31.**

| | translational degrees of freedom | rotational degrees of freedom | vibrational degrees of freedom |
|---|---|---|---|
| a) | 3 | 3 | 9 |
| b) | 3 | 2 | 4 |
| c) | 3 | 3 | 30 |
| d) | 3 | 2 | 7 |
| e) | 3 | 3 | 6 |
| f) | 3 | 3 | 21 |

**10-32.** The dipole moment changes during the motion in all the modes except $\nu_1$, and hence they all are infrared active except $\nu_1$.

**10-33.** Use the equation (p. 463)

$$\bar{\nu}_{00} = \bar{\nu}_{el,el} - \left[\frac{1}{2}\bar{\nu}_e' - \frac{1}{4}\bar{x}_e'\bar{\nu}_e'\right] + \left[\frac{1}{2}\nu_e'' - \frac{1}{4}\bar{x}_e''\bar{\nu}_e''\right]$$

$$\bar{\nu}_{el,el} = 64751.4 \text{ cm}^{-1} - (752.7 \text{ cm}^{-1}) + (1081.7 \text{ cm}^{-1})$$
$$= 65080.4 \text{ cm}^{-1}$$

**10-34.** Use Eq. 10-52,

$$\bar{E} = \bar{\nu}_{el} + \bar{\nu}_e(n + 1/2) - \bar{x}_e\bar{\nu}_e(n + 1/2)^2$$

in the form

$$\bar{E}'/\text{cm}^{-1} = 65080.43 + 1514.10(n' + 1/2) - 17.40(n' + 1/2)^2$$

$$\bar{E}''/\text{cm}^{-1} = 2170.21(n'' + 1/2) - 13.46(n'' + 1/2)^2$$

Let n' and n" equal zero to get

$$\bar{E}'(\text{zero-point}) = 65833.13 \text{ cm}^{-1}$$

$$\bar{E}''(\text{zero-point}) = 1081.74 \text{ cm}^{-1}$$

$$\bar{\nu}_{obs}(0 \to 0) = \bar{E}'(n' = 0) - \bar{E}''(n'' = 0) = 64751.4 \text{ cm}^{-1}$$

$$\bar{\nu}_{obs}(0 \to 1) = \bar{E}'(n' = 1) - \bar{E}''(n'' = 0) = 67312.3\text{cm}^{-1} - 1081.74 \text{ cm}^{-1} = 66230.6 \text{ cm}^{-1}$$

$$\bar{\nu}_{obs}(0 \to 2) = \bar{E}'(n' = 2) - \bar{E}''(n'' = 0) = 68756.7 \text{ cm}^{-1} - 1081.74 \text{ cm}^{-1} = 67674.9 \text{ cm}^{-1}$$

$$\bar{\nu}_{obs}(0 \to 3) = \bar{E}'(n' = 3) - \bar{E}''(n'' = 0) = 70166.3 \text{ cm}^{-1} - 1081.74 \text{ cm}^{-1} = 69084.5 \text{ cm}^{-1}$$

$$\bar{\nu}_{obs}(0 \to 4) = 71541.1 \text{ cm}^{-1} - 1081.74 \text{ cm}^{-1} = 70459.4 \text{ cm}^{-1}$$

**10-35.** Use the fact that

$$\bar{B}_{n+1} - \bar{B}_n = -\bar{\alpha}_e$$

| $n$ | $\bar{\alpha}_e/\text{cm}^{-1}$ |
|---|---|
| 0 | 0.00592 |
| 1 | 0.00592 |
| 2 | 0.00590 |

Given $\bar{\alpha}_e = 0.00592 \text{ cm}^{-1}$, we find that

$$\bar{B}_e = 0.82004 \text{ cm}^{-1}\ (n = 0);\ 0.82004 \text{ cm}^{-1}\ (n = 1);\ 0.82004 \text{ cm}^{-1}\ (n = 2)$$

**10-36.**

$$\bar{\nu}_R = \bar{\nu}_{el,vib} + 2\bar{B}' + (3\bar{B}' - \bar{B}'')J + (\bar{B}' - \bar{B}'')J^2 \qquad J = 0,1,2,\ldots$$

$$\bar{\nu}_P = \bar{\nu}_{el,vib} - (\bar{B}' + \bar{B}'')J + (\bar{B}' - \bar{B}'')J^2 \qquad J = 1,2,3,\ldots$$

Introducing values of $\bar{B}'$ and $\bar{B}''$, we obtain

$$(\bar{\nu}_R - \bar{\nu}_{el,vib})/\text{cm}^{-1} = 19.6 + 18.15\,J - 1.45\,J^2 \qquad J = 0,1,2,\ldots$$

$$(\bar{\nu}_P - \bar{\nu}_{el,vib})/\text{cm}^{-1} = -\,21.05\,J - 1.45\,J^2 \qquad J = 1,2,3,\ldots$$

The Fortrat diagram is

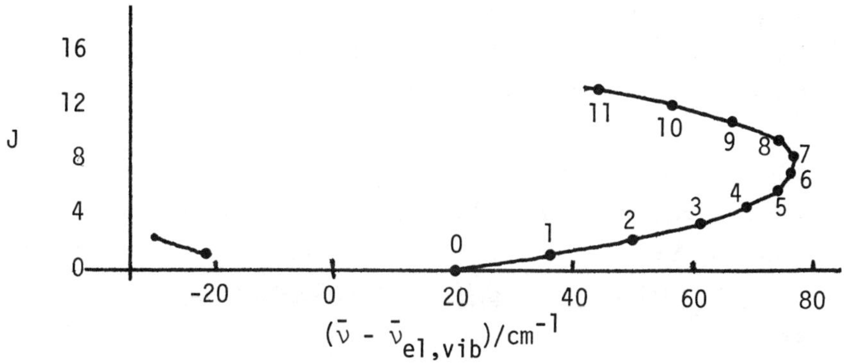


The relative frequency at the band head is about 77 cm$^{-1}$.

**10-37.** Substitute Eq. 10-59 into Eq. 10-57 to obtain

$$-\frac{\hbar^2}{2M}(\nabla_A^2 + \nabla_B^2)\psi_{el}(\mathbf{r}_1,\mathbf{r}_2;R)\psi_n(\mathbf{R}_A,\mathbf{R}_B) + \hat{H}_{el}\psi_{el}(\mathbf{r}_1,\mathbf{r}_2;R)\psi_n(\mathbf{R}_A,\mathbf{R}_B) = E\psi_{el}(\mathbf{r}_1,\mathbf{r}_2;R)\psi_n(\mathbf{R}_A,\mathbf{R}_B)$$

We now assume that $\psi_{el}(\mathbf{r}_1,\mathbf{r}_2;R)$ varies sufficiently smoothly with R that we can write

$$(\nabla_A^2 + \nabla_B^2)\psi_{el}(\mathbf{r}_1,\mathbf{r}_2;R)\psi_n(\mathbf{R}_A,\mathbf{R}_B) \approx \psi_{el}(\mathbf{r}_1,\mathbf{r}_2;R)(\nabla_A^2 + \nabla_B^2)\psi_n(\mathbf{R}_A,\mathbf{R}_B)$$

in which case we recover Eq. 10-60.

**10-38.** The quantity $E_{el}(R)$ is the same for $H_2$, HD, and $D_2$ under the Born-Oppenheimer approximation. Because the equilibrium nuclear separation is determined by the minimum in $E_{el}(R)$ and the force constant is determined by the curvature of $E_{el}(R)$ at the minimum, these two quantities are predicted to be the same for $H_2$, HD and $D_2$ under the Born-Oppenheimer approximation. Because $\nu_{obs} = (2\pi)^{-1}(k/\mu)^{1/2}$ and $B = h/8\pi^2\mu R^2_{min}$, these are also predicted to be the same.

**10-39.** Under the Born-Oppenheimer approximation, the force constant of $H^{35}Cl$ is equal to that of $D^{35}Cl$. Thus we use

$$k = (2\pi c\bar{\nu}_{HCl})^2\mu_{HCl} = (2\pi c\bar{\nu}_{DCl})^2\mu_{DCl}$$

or

$$\frac{\bar{\nu}_{DCl}}{\bar{\nu}_{HCl}} = \left[\frac{\mu_{HCl}}{\mu_{DCl}}\right]^{1/2} = \left[\frac{35/36}{70/37}\right]^{1/2} = 0.717$$

$$\bar{\nu}_{DCl} = (0.717)(2990.0\ \text{cm}^{-1}) = 2143.4\ \text{cm}^{-1}$$

**10-40.** Use the equations (see Problem 10-39)

$$\frac{\bar{\nu}_{HD}}{\bar{\nu}_{H_2}} = \left[\frac{\mu_{H_2}}{\mu_{HD}}\right]^{1/2} = \left[\frac{1/2}{2/3}\right]^{1/2} = 0.866$$

and

$$\frac{\bar{\nu}_{D_2}}{\bar{\nu}_{H_2}} = \left[\frac{\mu_{H_2}}{\mu_{D_2}}\right]^{1/2} = \left[\frac{1/2}{4/4}\right]^{1/2} = 0.707$$

Thus

$$\bar{\nu}_{HD} = (0.866)(4395.20\ \text{cm}^{-1}) = 3806\ \text{cm}^{-1}$$

$$\bar{\nu}_{D_2} = (0.707)(4395.20\ \text{cm}^{-1}) = 3107\ \text{cm}^{-1}$$

Because

$$B = \frac{h}{8\pi^2\mu R^2_{min}}$$

and $R_{min}$ is the same for $H_2$, HD and $D_2$ under the Born-Oppenheimer approximation, we have

$$\frac{B_{HD}}{B_{H_2}} = \frac{\mu_{H_2}}{\mu_{HD}} = \frac{3}{4}$$

and

$$\frac{B_{D_2}}{B_{H_2}} = \frac{\mu_{H_2}}{\mu_{D_2}} = \frac{1}{2}$$

Thus

$$B_{HD} = (0.75)(59.31\ \text{cm}^{-1}) = 44.48\ \text{cm}^{-1}$$

$$B_{D_2} = (0.50)(59.31\ \text{cm}^{-1}) = 29.66\ \text{cm}^{-1}$$

**10-41.** Introduce

$$X = \frac{Mx_1 + Mx_2}{2M} = \frac{1}{2}x_1 + \frac{1}{2}x_2$$

$$x = x_1 - x_2$$

Now, using the chain rule for partial differentiation

$$\frac{\partial^2 f}{\partial x_1^2} = \frac{\partial}{\partial x_1}\left[\frac{\partial f}{\partial X}\frac{\partial X}{\partial x_1} + \frac{\partial f}{\partial x}\frac{\partial x}{\partial x_1}\right]$$

$$= \frac{\partial}{\partial x_1}\left[\frac{1}{2}\frac{\partial f}{\partial X} + \frac{\partial f}{\partial x}\right]$$

$$= \frac{1}{2}\left[\frac{\partial X}{\partial x_1}\frac{\partial^2 f}{\partial X^2} + \frac{\partial x}{\partial x_1}\frac{\partial^2 f}{\partial X\,\partial x}\right] + \left[\frac{\partial X}{\partial x_1}\frac{\partial^2 f}{\partial X\,\partial x} + \frac{\partial x}{\partial x_1}\frac{\partial^2 f}{\partial x^2}\right]$$

$$= \frac{1}{4}\frac{\partial^2 f}{\partial X^2} + \frac{1}{2}\frac{\partial^2 f}{\partial X\,\partial x} + \frac{1}{2}\frac{\partial^2 f}{\partial X\,\partial x} + \frac{\partial^2 f}{\partial x^2}$$

Similarly

$$\frac{\partial^2 f}{\partial x_2^2} = \frac{1}{4}\frac{\partial^2 f}{\partial X^2} - \frac{\partial^2 f}{\partial X\,\partial x} + \frac{\partial^2 f}{\partial x^2}$$

Thus

$$\frac{\partial^2}{\partial x_1^2} + \frac{\partial^2}{\partial x_2^2} = \frac{1}{2}\frac{\partial^2}{\partial X^2} + 2\frac{\partial^2}{\partial x^2}$$

and

$$-\frac{\hbar^2}{2M}\left[\frac{\partial^2}{\partial x_1^2} + \frac{\partial^2}{\partial x_2^2}\right] = -\frac{\hbar^2}{4M}\frac{\partial^2}{\partial X^2} - \frac{\hbar^2}{M}\frac{\partial^2}{\partial x^2}$$

$$= -\frac{\hbar^2}{2M_{tot}}\frac{\partial^2}{\partial X^2} - \frac{\hbar^2}{2\mu}\frac{\partial^2}{\partial x^2}$$

where we have used the facts that $M_{tot} = 2M$ and $\mu = M/2$.

Eq. 10-60 now becomes

$$\left[-\frac{\hbar^2}{2M_{tot}}\frac{\partial^2}{\partial X^2} - \frac{\hbar^2}{2\mu}\frac{\partial^2}{\partial x^2}\right]\psi(X,x) + E_{el}(x)\psi(X,x)$$

$$= E\psi(X,x)$$

or

$$\left[-\frac{\hbar^2}{2M_{tot}}\frac{\partial^2}{\partial X^2} - \frac{\hbar^2}{2\mu}\frac{\partial^2}{\partial x^2} + E_{el}(x)\right]\psi(X,x)$$

$$= E\psi(X,x)$$

The Hamiltonian operator is separable (Section 6-2), and so the problem can be separated into one involving only the center-of-mass coordinate and one involving only the relative coordinate.

**10-42.** Note that $\psi_{vib}(R) = S_{vib}(R - R_e)/R$, and although the volume element of $\psi_{vib}(R)$ is $R^2dR$, that of $S_{vib}(R)$ is simply $dR$. The orthogonality integral is

$$\int_{-\infty}^{\infty} S_{vib,n}(R)S_{vib,n'}(R)dR = \delta_{nn'}$$

The lower limit can be set as $-\infty$ rather than $-R_e$ ($R - R_e$ when $R = 0$) because $S_{vib}$ goes to zero rapidly as $R \to 0$.

**10-43.** For the $0 \to 1$ transition (Example 10-13),

$$I_{0\to1} \sim \int_0^{2\pi} d\phi \int_0^{\pi} d\theta \sin\theta\cos\theta \; Y_1^0(\theta,\phi)Y_0^0(\theta,\phi)$$

$$\sim 2\pi\int_0^{\pi} d\theta\sin\theta\cos\theta\left[\frac{3}{4\pi}\right]^{1/2}\cos\theta\left[\frac{1}{4\pi}\right]^{1/2}$$

$$\sim \frac{1}{\sqrt{3}}$$

$$I_{1\to2} \sim \int_0^{2\pi} d\phi \int_0^{\pi} d\theta\sin\theta\cos\theta \; Y_2^0(\theta,\phi)Y_1^0(\theta,\phi)$$

$$\sim 2\pi\int_0^\pi d\theta \sin\theta\cos\theta \left[\frac{5}{16\pi}\right]^{1/2}(3\cos^2-1)\left[\frac{3}{4\pi}\right]^{1/2}\cos\theta$$

$$\sim \frac{(15)^{1/2}}{4}\int_{-1}^{1}dx\ x^2(3x^2-1) = \frac{2}{\sqrt{15}}$$

$$\text{ratio} = \frac{I_{0\to1}}{I_{1\to2}} = \frac{\sqrt{5}}{2}$$

**10-44.** For the 0 → 1 transition (Example 10-14),

$$I_{0\to1} \sim \int_{-\infty}^{\infty}\psi_1(\xi)\,\xi\psi_0(\xi)d\xi$$

$$\sim \left[\frac{2\alpha}{\pi}\right]^{1/2}\int_{-\infty}^{\infty}\xi^2 e^{-\xi^2}d\xi = \frac{\alpha^{1/2}}{\sqrt{2}}$$

$$I_{1\to2} \sim \int_{-\infty}^{\infty}\psi_2(\xi)\,\xi\psi_1(\xi)d\xi$$

$$\sim \left[\frac{\alpha}{\pi}\right]^{1/2}\int_{-\infty}^{\infty}\xi^2(2\xi^2-1)e^{-\xi^2}d\xi = \alpha^{1/2}$$

$$\text{ratio} = \frac{I_{0\to1}}{I_{1\to2}} = \frac{1/\sqrt{2}}{1} = \frac{1}{\sqrt{2}}$$

**10-45.** From Eq. 10-103,

$$I = \int\psi'^{*}_{el}\mu_{el}\psi_{el}d\tau_e$$

Consider first the z component of $\mu_{el}$

$$I = \int\psi_{n'\ell'm'}(r,\theta,\phi)^{*}r\cos\theta\psi_{n\ell m}(r,\theta,\phi)d\tau$$

The angular integration gives

$$I \sim \int_0^{2\pi}d\phi\int_0^{\pi}d\theta\sin\theta\ Y_{\ell'}^{m'}(\theta,\phi)^{*}\cos\theta Y_{\ell}^{m}(\theta,\phi)$$

The integral over $\phi$ immediately shows that $m' = m$, and so we have that

$$I \sim \int_0^{\pi}d\theta\sin\theta\ P_{\ell'}^{|m|}(\cos\theta)\cos\theta P_{\ell}^{|m|}(\cos\theta)$$

$$\sim \int_{-1}^{1} dx P_{\ell'}^{|m|}(x) x P_{\ell}^{|m|}(x)$$

To evaluate this integral, we need the relation that is given in Problem 6-21

$$(2\ell + 1)x\ P_{\ell}^{|m|}(x)$$
$$= (\ell - |m| + 1)P_{\ell+1}^{|m|}(x) + (\ell + |m|)P_{\ell-1}^{|m|}(x)$$

Using this relation, we have

$$I \sim \frac{(\ell - |m| + 1)}{(2\ell + 1)} \int_{-1}^{1} dx P_{\ell'}^{|m|}(x) P_{\ell+1}^{|m|}(x)$$
$$+ \frac{(\ell + |m|)}{(2\ell + 1)} \int_{-1}^{1} dx P_{\ell'}^{|m|}(x) P_{\ell-1}^{|m|}(x)$$

Because of the orthogonality of the $P_{\ell}^{|m|}(x)$ (Eq. 6-74), we see that $I = 0$ unless $\ell' = \ell + 1$ or $\ell' = \ell - 1$. Thus the selection rules in the spectrum of atomic hydrogen are $m' = m$ ($\Delta m = 0$) and $\ell' = \ell \pm 1$ ($\Delta\ell = \pm 1$).

The x and y components of $\mu_{el}$ give integrals over $\phi$ like

$$\int_{0}^{2\pi} d\phi e^{-im'\phi} \begin{Bmatrix} \cos\phi \\ \sin\phi \end{Bmatrix} e^{im\phi}$$

Write both $\cos\phi$ $\sin\phi$ as

$$\begin{Bmatrix} \cos\phi \\ \sin\phi \end{Bmatrix} = \begin{Bmatrix} \dfrac{e^{i\phi} + e^{-i\phi}}{2} \\ \dfrac{e^{i\phi} - e^{-i\phi}}{2i} \end{Bmatrix}$$

to get

$$\int_{0}^{2\pi} d\phi e^{i(m-m'\pm 1)\phi}$$

This integral vanishes unless $m - m' \pm 1 = 0$, or unless

$$m' = m \pm 1$$

Thus $\Delta m = \pm 1$ for $\mu_{x,el}$ and $\mu_{y,el}$.

# APPENDIX B
# COMPLEX NUMBERS

**B-1.**

(a) $(2 - i)^3 = 8 - 12i + 6i^2 - i^3$

$= 8 - 12i - 6 + i$

$= 2 - 11i$

(b) $e^{\pi i/2} = \cos(\pi/2) + i\sin(\pi/2) = i$

(c) $e^{-2+i\pi/2} = e^{-2}(\cos(\pi/2) + i\sin(\pi/2)) = ie^{-2}$

(d) $(\sqrt{2} + 2i)e^{-i\pi/2} = (\sqrt{2} + 2i)(\cos(\pi/2) - i\sin(\pi/2)$

$= (\sqrt{2} + 2i)(-i) = 2 - i\sqrt{2}$

(e) $(1 - i)(2 + 3i)(-2 + i)$

$= (1 - i)(-4 - 6i + 2i - 3)$

$= (1 - i)(-7 - 4i)$

$= - 7 + 7i - 4i - 4$

$= - 11 + 3i$

**B-2.** $z = x + iy$

$z^* = x - iy$

$z + z^* = 2x = 2\text{Re } z$

$z - z^* = 2iy = 2\text{Im } z$

**B-3.**

(a) $\text{Re } z^* = \text{Re}(x - 2iy) = x$

(b) $\text{Re } z^2 = \text{Re}(x^2 - 4y^2 + 4ixy) = x^2 - 4y^2$

(c) $\text{Im } z^2 = \text{Im}(x^2 - 4y^2 + 4ixy) = 4xy$

(d) $\text{Re } zz^* = \text{Re}(x^2 + 4y^2) = x^2 + 4y^2$

(e) $\text{Im } zz^* = \text{Im}(x^2 + 4y^2) = 0$

**B-4.**

(a) $|z| = 6, \ \phi = 90°$

(b) $|z| = (16 + 2)^{1/2} = (18)^{1/2} = 3\sqrt{2}$

$\phi = \tan^{-1}(-\sqrt{2}/4) = -19.5°$ (fourth quadrant)

(c) $|z| = (1 + 4)^{1/2} = \sqrt{5}$

$\phi = \tan^{-1}(-2/-1) = 243.4°$ (third quadrant)

(d) $|z| = (\pi^2 + e^2)^{1/2}$

$\phi = \tan^{-1}(e/\pi) = 40.9°$

(e) $|z| = 3$

$\phi = 180°$

(f) $|z| = (5 + 4)^{1/2} = 3$

$\phi = \tan^{-1}(-2/\sqrt{5}) = -41.8°$

**B-5.** $e^{i\pi} = \cos\pi + i\sin\pi = -1$

Although e and π are irrational numbers, $e^{i\pi} = -1$.

**B-6.** Start with Euler's formula

$$e^{i\theta} = \cos\theta + i\sin\theta$$

$$e^{-i\theta} = \cos(-\theta) + i\sin(-\theta) = \cos\theta - i\sin\theta$$

Thus

$$e^{i\theta} + e^{-i\theta} = 2\cos\theta$$

and

$$e^{i\theta} - e^{-i\theta} = 2i\sin\theta$$

**B-7.** Start with

$$z = r(\cos\theta + i\sin\theta) = re^{i\theta}$$

Thus

$$z^n = r^n e^{in\theta} = r^n(\cos n\theta + i\sin n\theta)$$

and

$$e^{in\theta} = (\cos\theta + i\sin\theta)^n = \cos n\theta + i\sin n\theta$$

**B-8.** Start with (Problem B-7)

$$(\cos\theta + i\sin\theta)^n = \cos n\theta + i\sin n\theta$$

Let n = 2 to obtain

$$\cos^2\theta - \sin^2\theta + 2i\cos\theta\sin\theta = \cos 2\theta + i\sin 2\theta$$

Equate the real and imaginary parts

$$\cos 2\theta = \cos^2\theta - \sin^2\theta$$

$$\sin 2\theta = 2\sin\theta\cos\theta$$

Let n = 3

$$\cos^3 - 3\cos\theta\sin^2\theta + i(3\cos^2\theta\sin\theta - \sin^3\theta)$$

$$= \cos 3\theta + i\sin 3\theta$$

Equate real and imaginary parts to obtain

$$\cos 3\theta = \cos^3\theta - 3\cos\theta\sin^2\theta$$
$$= 4\cos^3\theta - 3\cos\theta$$
$$\sin 3\theta = 3\cos^2\theta\sin\theta - \sin^3\theta$$
$$= 3\sin\theta - 4\sin^3\theta$$

**B-9.**

$$\int_0^{2\pi} d\phi e^{im\phi} = \int_0^{2\pi} d\phi \cos m\phi + i\int_0^{2\pi} d\phi \sin m\phi = 0 \qquad m \neq 0$$

because both integrals are over complete cycles of $\cos m\phi$ and $\sin m\phi$ if $m \neq 0$. If $m = 0$, we have

$$\int_0^{2\pi} d\phi = 2\pi \qquad\qquad m = 0$$

Thus

$$\int_0^{2\pi} d\phi \Phi_m(\phi) = \frac{1}{\sqrt{2\pi}}\int_0^{2\pi} d\phi e^{im\phi} = \sqrt{2\pi}\ \delta_{m0}$$

Now

$$\int_0^{2\pi} d\phi \Phi_m(\phi)^* \Phi_n(\phi) = \frac{1}{2\pi}\int_0^{2\pi} d\phi e^{i(n-m)\phi} \begin{array}{l} = 1 \text{ if } n = m \\ = 0 \text{ if } n \neq m \end{array}$$

**B-10.** Start with

$$f(\theta) = \ln(\cos\theta + i\sin\theta)$$

Differentiate to obtain

$$\frac{df}{d\theta} = \frac{-\sin\theta + i\cos\theta}{\cos\theta + i\sin\theta} = \frac{i(\cos\theta + i\sin\theta)}{\cos\theta + i\sin\theta} = i$$

Integrate this to obtain

$$f(\theta) = i\theta + c = \ln(\cos\theta + i\sin\theta)$$

Let $\theta = 0$ to find that $c = \ln(1) = 0$, and then take the antilogarithm and write

$$\cos\theta + i\sin\theta = e^{i\theta}$$